T0278278

LONG ISLAND
STATE PARKS

LONG ISLAND
STATE PARKS

A HISTORY FROM JONES BEACH TO MONTAUK

KRISTEN MATEJKA

THE
History
PRESS

Published by The History Press
Charleston, SC
www.historypress.com

Copyright © 2024 by Kristen Matejka
All rights reserved

Front cover, top (left to right): Kitchen staff at Kings Park Psychiatric Center, 1924. *Kings Park Heritage Museum.* Jones Beach State Park shortly after it opened to the public. *Long Island State Parks.* Young residents of the Society of St. Johnland, 1935. *Kings Park Heritage Museum.*
Front cover, bottom: Bayard Cutting Arboretum. *Author's collection.*

First published 2024

Manufactured in the United States

ISBN 9781467157995

Library of Congress Control Number: 2024936793

Notice: The information in this book is true and complete to the best of our knowledge. It is offered without guarantee on the part of the author or The History Press. The author and The History Press disclaim all liability in connection with the use of this book.

All rights reserved. No part of this book may be reproduced or transmitted in any form whatsoever without prior written permission from the publisher except in the case of brief quotations embodied in critical articles and reviews.

This book is dedicated to my parents, Linda and Jerry Matejka.
Thank you.

CONTENTS

CONTENTS

ACKNOWLEDGEMENTS

Warmest thanks to all the people who walked this path with me and offered their encouragement, knowledge and enthusiasm along the way.

Anyone who has sat down and written a book can attest to the fact that it is equally one of the most challenging and most rewarding things a writer can do. Hours and hours of research and digging. Looking for clues and compiling facts. Finding the right people who know the answers or can point you in the right direction.

I am grateful for the historians who took the time to meet with me, opened their archives or shared their knowledge. Some of them were volunteers or employees at local historical societies and museums, while others were designated town historians. I also want to express my gratitude to our public libraries. Our libraries are a vast resource for Long Island history, and the history and reference librarians I encountered were eager to help me find information and historic images. Thank you.

Thanks to park employees at places such as Hempstead Lake State Park, Caleb Smith State Park, Wildwood State Park, Connetquot River State Park and others, who expressed a genuine interest in and knowledge about the individual parks where they work and readily shared their stories with me.

Special thanks to Mary Cascone, town historian, Town of Babylon Office of Historic Services; Cinda Lawrence, assistant to the president, Boys' Club of New York; Robert Hughes, Huntington town historian; and George Munkenbeck, Islip town historian, among many others. I met so many interesting people with a wealth of information and tales to share. It was my privilege.

I want to express my gratitude to the individuals who compiled information through the Livebrary system offered through my local library. This online tool is an excellent resource for researching historic newspapers and publications. Additionally, I am thankful for New York Heritage, a project of the Empire State Library Network, and the Library of Congress. These were invaluable in aiding my research.

I am especially thankful to Long Island State Parks and the New York State Office of Parks, Recreation and Historic Preservation. Specifically, I would like to thank John Williams, Long Island State Parks photographer, who shared images from the state archive and also shared my enthusiasm for digging around and finding information. As a photographer, he talked about how writing a good story is like getting a good photograph. You have to walk around it, look for different angles and find the right composition. He understood the challenge and the fun of being out there and figuring it out.

This book was written entirely without artificially generated intelligence, for which I am grateful. AI will never be able to duplicate the personal reward of formulating an idea and pursuing it. Meeting and corresponding with people and, most of all, physically visiting the parks to experience them firsthand to search for clues is something AI can never replicate.

Perhaps, most of all, I would like to thank J. Banks Smither, acquisitions editor at The History Press. I will never forget the complete sense of astonishment and excitement I had when I received his correspondence expressing interest in my book idea. What an incredible honor. Thank you for believing in me and guiding me through this process.

In addition to all the wonderful people I met who assisted along the way, those closest to me supported and motivated me to keep going.

Special thanks to my daughter Katie, who tolerated my long stretches of silence and seclusion while I wrote for countless hours. Thank you for going out to Montauk to serve as my "correspondent," taking photographs and reporting back from Shadmoor and Napeague.

In addition, Andrew Komarek, a longtime family friend, more than once said, in his own no-nonsense way, "Just do it," when I talked about the idea for this book during the COVID-19 pandemic. Repeatedly, he taunted me with phrases like, "Well, what are you doing anyway?" during that truly unprecedented and unsettling time during quarantine—until I sat down and started an outline.

Thank you to my parents, Linda and Jerry Matejka, who always had my back and are truly remarkable in their own right. I always admired my mother's incredible sense of detail, her inquisitiveness, her analytical way of thinking and her innate ability to carefully examine and think around a challenge. Special thanks to my father, who encouraged me to go off the beaten path to see the world and explore. My father helped me develop a sense of adventure, and during the writing of this book, he offered constant advice and encouragement, as he has always done.

I hope this book will inspire visitors to view our state parks on Long Island as more than just the wonderful places of leisure they are. They truly offer a snapshot back to a different period in Long Island's history. While many of the parks have been modified to accommodate large-scale public recreation, if you take the time, you can still discover traces of an earlier time. Enjoy!

INTRODUCTION

Every year, millions of people visit Long Island state parks without realizing that each park has a fascinating story to tell about the remarkable history of the area. These parks give visitors a glimpse back to an earlier time, before the rise of housing developments, shopping malls, highways and industry. They preserve fragments of Long Island's rich and storied past.

Some parks were once the vast estates New York's elite built during the Gilded Age at the turn of the twentieth century. Others were "hunting clubs," where wealthy members banded together to create sportsmen's retreats. One was a camp where disadvantaged children from the city could enjoy the country air. Others were former psychiatric care facilities where New York's most vulnerable could receive treatment outside the cramped and overcrowded city wards. One served as a training camp where soldiers prepared for battle in the first national army, and another was a military base that was established to protect the interests of the country. Some simply hold stories about the local community where they were established and the people who lived and worked there.

While many parks maintain the historical structures associated with their past, others offer only clues to their earlier use. Preservation was not necessarily the primary intent at the time many of the state parks were created. They were initially established to address the growing public need for outdoor spaces and recreation.

In the mid-1920s, the population of New York City was rapidly growing, and the state recognized the urgent necessity to provide the public with healthy places where they could breathe fresh air and engage in leisure activities. The use of air-conditioning was not prevalent back then, and the apartments and streets in New York City could become unbearably hot during summers.

In 1924, New York governor Alfred "Al" Smith supported legislation drafted by Robert Moses to establish the Long Island State Park Commission and the State Council on Parks. Under the proposed legislation, the commission would acquire vast areas of land to create parks for public use. Voters applauded the idea and approved a $15 million bond to put the park plan in action. Robert Moses was named president of the commission, and he would go on to hold a dozen related positions throughout the state during his career.

The few parks that existed at the time, including Niagara Falls (1885) and Bear Mountain (1913), were immensely popular. More parks were needed closer to New York's urban centers, and Long Island was a natural choice. At the start of the Long Island State Park Commission's work, Long Island was a very different place. It was sparsely developed, there were no major highways and most of the island comprised rural farmland with unpaved roads. Some New York City residents had already started to build small bungalow communities in waterfront areas of Long Island, while wealthy Manhattanites created lavish Gold Coast mansions along the North Shore to escape the city heat. There were resorts and clubs accessible by train or steamboat for those who could afford the trip.

In the first six years of its work, from 1924 to 1930, the Long Island State Park Commission acquired land for thirteen major parks and the land to construct major portions of the parkways leading to them. Today, there are twenty-seven major Long Island state parks and numerous more golf courses, nature preserves and historic sites operated by New York State. The vision to create a system of parks purely for the public's enjoyment was, from a public works perspective, simply astounding.

The initial Long Island State Park Commission acquisitions included Jones Beach State Park, Gilgo State Park, Belmont Lake State Park, Bethpage State Park, Captree State Park, Wildwood State Park, Orient Beach State Park, Heckscher State Park and Sunken Meadow State Park. In addition, the parks commission acquired additional acreage for Fire Island State Park (later Robert Moses State Park), as well as an easement from the City of New York to create Valley Stream State Park, Hempstead Lake State Park and Massapequa State Park Preserve.

During this period, the number of cars on the road increased from eight million in 1918 to twenty-three million in 1929. The automobile was fast becoming the desired means of transportation. America had become mobile. The parkways were created with a commonality in layout and design. They all feature the same rustic stone overpasses and are all lined with trees and grass to create a meandering band of interconnected, scenic roadways. The intent was to make the parkways, or the "way to the parks," feel like parks as well. Their appearance is in sharp contrast to that of the other major commercial roadways later built on Long Island.

Some parkways were never completed, such as one that was to bring visitors to Caumsett State Park. Plans at one time also included parkways along Fire Island and the North Shore of Long Island. There were even talks of bringing parkways out to Orient State Park, Hither Hills and Montauk Point State Park. While the original vision was for a grand loop of attractive parkways to lead to all the parks, this was not feasible for some of the later acquisitions.

In 1928, New York governor Alfred Smith left office to run for the United States presidency. New York governor Franklin Delano Roosevelt, who served until 1932, succeeded him. Roosevelt continued to support the efforts of the Long Island State Park Commission, acquiring state parkland for the public and making improvements to existing parkland. In 1932, Roosevelt was elected president of the United States amid the Great Depression. He is remembered for rallying the nation's citizens in this turbulent time and creating the Works Progress Administration (WPA), which put hundreds of thousands of people back to work on massive public projects as part of his New Deal. On Long Island, some of the park facilities were constructed or expanded using New Deal money. The WPA conducted improvement projects at Sunken Meadow State Park, Belmont Lake State Park, Wildwood State Park, Orient Point State Park and Hither Hills State Park, as well as Bethpage State Park, home of the famous Bethpage Black golf course. The WPA also built the imposing twelve-story Building 93 at Nissequogue River State Park when that park was a psychiatric hospital.

Long Island State Park Commission continued to acquire tracts of land for public recreation where it could. On Long Island, the landscape was changing rapidly after the recovery of the economy in the 1940s and 1950s. Large suburban housing developments began sprouting up, such as the first mass-produced suburb in Levittown, covering what had largely been farmland, woodland and open plains.

During that time, Robert Moses pleaded with successive governors and Long Island officials to "obtain more land for recreation while there is still time," according to a 1958 article in the *New York Times*. That statement was actually quite prescient. In that same article, the park commissioner is quoted as saying that more parkland for recreation should be created "before demon realtors get in with bulldozers and tear everything to pieces." Today, Long Island is the most densely populated island in the United States.

Robert Caro, who, in in 1975, wrote the Pulitzer Prize–winning book *The Power Broker*, which is the definitive biography on Robert Moses, said:

> *As for the parks he created, fly over New York in the year 1999, and the two thousand acres of Brookhaven Park and four thousand acres of Connetquot on Long Island, and the twenty-one thousand other acres of parks that Robert Moses wrested away from the developer's bulldozer to insure that the people of New York would always have green space that will still be green is a tribute to his foresight.*

Between the 1920s and 1960s, New York State acquired more than 2.5 million acres of parkland and built 658 playgrounds across the five boroughs of New York, Long Island and Westchester. During this time, Robert Moses held many other state posts, including chairman of the New York State Council of Parks, commissioner of the New York City Department of Parks and commissioner of the New York City Planning Commission, among many others.

It is important to clarify that this book does not focus on Robert Moses or any of the other massive and numerous projects he was associated with while leading other New York commissions. Moses was a complex figure whose ambitions forever altered the landscape and neighborhoods in the New York metro area. Rather, this book sheds light on the natural beauty and historical heritage that the Long Island State Park Commission preserved in perpetuity.

Even after Moses relinquished all of the city and state posts he held, the New York State Office of Parks, Recreation and Historic Preservation, created in 1970, continued to preserve open space for recreation in what are now the vast, sprawling suburbs of modern-day Long Island. However, a fiscal crisis in in the 1970s prevented the broad-scale acquisitions seen in the decades earlier. From 1977 until 1995, there was little state funding available to acquire new state parks. Then from 1995 to 2006, New York State governor George Pataki and Park Commissioner Bernadette Castro picked up the ball and added nine new parks on Long Island during their tenure, although subsequent budget restraints did not allow for immediate improvement on much of the parkland. The point was that the land was preserved.

The post-1995 park acquisitions include Cold Spring Harbor State Park, Trail View State Park, Nissequogue River State Park, Hallock State Park Preserve, Shadmoor State Park, Brentwood State Park, Camp Hero State Park, Amsterdam Beach State Park and Sag Harbor State Golf Course.

The Long Island State Park Commission gained much of the land for Long Island's parks through private sale, dedication by municipalities, donation by wealthy benefactors or outright acquisition. While the state's intention was to create vast areas of open space for the public's enjoyment, Long Island State Parks also hold secrets of Long Island's colorful and intriguing history.

Visit today, and you will often find remnants and hints to the parks' former use. Some parks offer an immersion into the natural world, showcasing woodlands and waterways that existed when Native people roamed these lands. Others hold local stories of early settlers and patentees, while others existed as places of great national significance.

Each park in Long Island's state park system has a unique tale to tell about the evolution of Long Island and the growth of the country and society in general. These stories include chronicles of wartime and peacetime; accounts of great wealth and simultaneous public need; and times of extensive cultural transformation.

They are time capsules from the past. If you observe closely, you can still find remnants of long ago and imagine what these lands were like back then.

This book explores the fascinating hidden stories behind Long Island State Parks before they were parks.

1

BAYARD CUTTING ARBORETUM STATE PARK

Bayard Cutting Arboretum State Park in Great River was once a grand country estate called Westbrook, built for William Bayard Cutting and his wife, Olivia Peyton Murray, in 1886.

Bayard Cutting made his fortune in railroads, working alongside his grandfather Robert Bayard in the management of the St. Louis, Alton and Terre Haute Railroad. He later went on to manage the Illinois Central Railroad and the Southern Pacific line and later expanded his interests to include financing and real estate development. Always ambitious, Bayard Cutting was most notably instrumental in the development of the South Brooklyn Waterfront. Its connection to the railroad and the ferry system enabled warehousing and shipping operations to develop. At one point, he even worked to have the Ambrose Channel broadened to allow for shipping in and out of the Port of New York, according to the official history of the park.

Bayard Cutting loved the outdoors, hunting and gardening. Therefore, in the early 1880s, he purchased land from George L. Lorillard on the east bank of the Connetquot River and along West Brook in order to create a magnificent estate. The property was previously a horse farm and game preserve.

Many are familiar with the Gold Coast of Long Island, where mansions of European caliper were constructed high on the bluffs of Long Island's North Shore. There, New York City's elite and the highest echelon of society sought to portray themselves as "American royalty," striving to outdo each other in terms of opulence and grandeur.

However, there were several mansions built in the South Shore area. The nearby South Side Sportsmen's Club likely drew Bayard Cutting to the area,

Historic Bayard Cutting estate. *East Islip Historical Society.*

since it had become popular for high-society men to seek rustic getaways and outdoor activities, such as hunting and fishing, to escape the demands of their everyday wealth building. Other "captains of industry," such as William K. Vanderbilt and Frederick Bourne, built vast and impressive estates on this part of the island as well.

When visiting the current state park, guests notice the most prominent feature is the many-gabled, sixty-room home Bayard Cutting had constructed on the property. New York City architect Charles C. Haight designed it in the Queen Anne style, with a timber and stucco finish and a wood-shingled roof that gives it a rustic, almost storybook, look. Stepping into the main hall of the mansion, guests are greeted by handcrafted woodwork and heavy paneling imported from England. The home features many magnificent fireplaces (including a mantel from a fourteenth-century French chateau) and medieval stained-glass windows designed by Louis Comfort Tiffany. The fireplace in the breakfast room arch is also a piece made by Tiffany. Much of the original décor remains, including an intricately carved dining room set and ornamental tapestries.

After the construction of the mansion, Bayard Cutting wanted the grounds to be as impressive as the house. To achieve this, he hired Frederick Law

Olmsted, a well-known landscape architect who also designed Central Park in New York City. Olmsted was known for his artistic approach to landscape design, which involved preserving the natural beauty of the land while incorporating fine specimens of trees from all around the world. The goal was not to create a formal English-style garden; rather, it was to produce a more rustic and less structured, or "English Naturalistic," park.

In addition to the fantastic arboretum Olmsted created, the grounds at Westbrook were used as a dairy farm, with a herd of Jersey cows producing bottled milk for sale in the local community. One of the original barns remains in the farm area of the current arboretum and is used as a visitor center.

The primary residence of Bayard Cutting and his wife, Olivia Peyton Murray Cutting, was located in New York City, since many of Bayard Cutting's business interests were in and around the city. The couple had four children, William, Justine, Bronson and Olivia, and the family enjoyed the estate for many years as a country, or summer, residence. Bayard Cutting was also a philanthropist. He was involved with improving the quality of urban housing in New York City and served on the boards of many religious, educational and charitable organizations, including the New York Botanical Garden, where his later avid interest in gardening and horticulture may have developed.

After Bayard's death, his wife, Olivia, maintained the estate and continued the family's philanthropic pursuits, creating grants and scholarships, along with donating generously to social relief funds. She also developed her own passion for public spaces and gardens, providing support for New York City parkland improvements. While the couple's four adult children enjoyed a comfortable life, philanthropy and public service were still a large part of the family dynamic. Unfortunately, tragedy struck down Bayard Cutting's two sons early in their lives. William Jr. suffered from tuberculosis, and after attending Harvard and serving in a British diplomatic role, he passed away from his affliction at the age of thirty-one. In addition, Bronson, who became a newspaper publisher, a captain during World War I and, later, a Republican U.S. senator, died in a plane crash in 1935 at the age of forty-seven. Justine went on to become an accomplished pianist and lived a long and productive life, passing away in 1975 at the age of ninety-six. Olivia became her mother's primary caregiver after her father's death when she was twenty years old. She married Henry James and became Mrs. Olivia Cutting James, according to the park's brief biographies of the family.

Olivia Cutting James and her mother were of the same mind when it came to conserving the estate for future generations to enjoy, as Bayard

Mrs. Cutting, Bronson, William Jr., Olivia, Mr. Cutting, Justine

William Bayard Cutting and family posing for a portrait. *East Islip Historical Society.*

Cutting may have wished. In 1936, Mrs. Cutting James generously donated 200 acres of the estate to New York Parks Commission, and in 1938, she donated another 441 acres in memory of her father. The family continued to live there until her mother passed away in the mansion in 1949. Shortly after, in 1952, Mrs. Cutting James officially handed the estate over to New York State, at which time, she established a trust to fund the maintenance of the property. A key stipulation in the agreement was that the state should maintain the property for the public's use and enjoyment. The bequest left by the family enabled the state to create visitor amenities and acquire an additional 50 acres, including the dairy.

The deed to the state expressly stipulates that the land and its plantings be preserved in perpetuity. The mission statement of the arboretum is to "provide an oasis of beauty and of quiet for the pleasure, rest and refreshment of those who delight in outdoor beauty, and to bring about a greater appreciation and understanding of the value and importance of informal plantings." Visitors may notice that the grounds have a gentle, meandering feel. Trails wind past flowering bushes and indigenous plantings and then, all at once, open into dramatic treed landscapes that reflect the "informal" planting style. The grounds are filled with firs, spruces, pines, hemlocks, cedars, cypresses and other conifers, some native and some rare and exotic. There are open fields and meadows, an expansive back lawn overlooking the river, marshland and wildflower gardens. The extensive plantings include rhododendrons, azaleas, hollies, oaks and a daffodil field at the far edge, near the water. The Old Pinetum on the property is especially noteworthy. Many of the trees there are well over one hundred years old, although sadly, many were lost during Hurricane Gloria in 1985.

Today, the grand mansion and 691 acres of rolling hills, trails, gardens and rare trees remain. Bayard Cutting Arboretum tells visitors not to think of this as merely a park; rather, it is a "museum of trees." In 1973, the site was listed in the National Register of Historic Places.

Some not-to-be-missed spots include the Woodland Garden, Paradise Island Walk, Four Seasons Walking Garden and the Conifer Garden, to name a few. Visitors can cross a footbridge and enjoy the vista from the rustic teahouse on Breezy Island. The original teahouse was constructed in the 1890s for Mrs. Bayard Cutting, but it developed structural issues and had to be torn down. It was later re-created based on the original architectural design. At the same time, a project was undertaken to rid the island of invasive bamboo and plant a garden of native plants that are more beneficial to the local ecosystem, including milkweed, northern highbush blueberry, common rush and more.

The public is invited to wander the grounds and admire the park's natural beauty. Tours of the mansion are available during regular hours, and the charming Hidden Oak Café, located inside the manor house, offers breathtaking views of the expansive yard and river while guests enjoy freshly made sandwiches, snacks, soups and pies. A traditional "Victorian tea" is available, by reservation, in the main house. There is plenty of indoor and outdoor seating on the back veranda and lawn.

In addition to providing a peaceful escape, the site acts as an educational resource. Signage is located throughout, and many trees are labeled so that visitors can gain an appreciation for their characteristics. Special programs teach the community about horticultural practices and sustainable farming.

2

BELMONT LAKE STATE PARK

Belmont Lake State Park, located in North Babylon, was once the residence of August Belmont. Born in Prussia (now Germany), Belmont was not particularly wealthy when he started his career. However, he managed to cross paths with the right people and worked his way up to become a financier and diplomat. Later, he established his own banking firm, August Belmont and Company, and became an avid horse breeder. The name Belmont is well known to anyone interested in American horseracing.

Early in Belmont's career, the internationally known Rothschild family mentored him. He worked in their Frankfurt, Germany office before he moved to New York in 1837. There, he established himself as the Rothschilds' United States agent, handling much of their banking interests.

The Rothschild family was a famous European banking dynasty that had far-reaching connections and wielded great influence in world events and politics at that time. In addition to banking, their interests included real estate, mining, finance and business venture development. Belmont was able to use his affiliation with the Rothschilds to build his great wealth in the United States, which later placed him squarely within the ranks of Gilded Age society.

As the wealthy did at that time, Belmont proceeded to build himself a twenty-four-room mansion on 1,100 acres of what is now the site of Belmont State Park on Long Island. Called the Belmont Nursery Stud Farms, the estate was built by Belmont in 1868 to accommodate his growing equestrian interests. It included 50 acres of paddocks for the horses and 150 acres of

Original August Belmont estate. *Long Island State Parks.*

grassland. In total, there were once thirty buildings on the site, including a conservatory, an icehouse, two silos, a blacksmith shop, a small gristmill and kennels for dogs. There was also a lake, a boathouse and squash courts. Five hundred acres were cultivated for produce, and five farm cottages housed farmworkers. Belmont also built an Episcopal chapel on the grounds for the workers. The church was later moved to Flushing, Queens, when the state demolished the original mansion and built the state park headquarters.

The buildings for the horses and equipment were located about a quarter-mile south of the main residence. There were bunks for over fifty grooms and a mile-long track where Belmont and his friends could watch the horses race. Belmont produced some of the most famous names in horseracing history in the nineteenth century, including Kingfisher, The Ill-Used and Fiddlesticks.

Belmont died at one of his other homes at 109 Fifth Avenue in New York City in 1890. His obituary noted that the cause of his death was pneumonia stemming from a cold he acquired at a horse show in Madison Square Garden, where he was serving as a judge. He apparently thought little of the cold at the time and continued to work himself until exhaustion set in and he collapsed.

After Belmont passed, he left the estate to his son August Belmont II, who went on to breed numerous horseracing legends, including the famous

Driveway to August Belmont estate, now the median on Southern State Parkway. *Long Island State Parks.*

Man o' War. In 1905, Belmont II and a group of partners founded Belmont Racetrack in Elmont. That racetrack on the Nassau–Queens border hosts the best of the best in horseracing at the annual Belmont Stakes. The Belmont Stakes is the third and most difficult leg of the prestigious annual Triple Crown tournament. A horse must first win the Kentucky Derby, then the Preakness and, finally, the Belmont Stakes before it can claim the Triple Crown title, given to only about two dozen horses in over one hundred years of the tournament's history.

Belmont II used the estate in North Babylon as a sportsman's retreat. He later bred cavalry animals for the United States Army. In 1918, during World War I, he agreed to lease 280 acres for army-training purposes. The army utilized the property to construct an airfield, hangars and barracks. The area was called Camp Damm and was used until 1919.

After Belmont II died in 1924, his widow, Eleanor Robson, a glamorous actress of the era, sold most of the estate to a developer who wanted to subdivide it into housing. The developer, Cadman H. Frederick, would go on

The 1879 tombstone for the Belmonts' "faithful dog" Robin. *Author's collection.*

to develop much of the land around the estate for housing. However, in 1926, he decided to sell the mansion, main farm buildings, lake and approximately 180 acres to the Long Island State Park Commission. Frederick sold the Belmont property to the commission for $97,000, which was quite a steal at the time, since the land and buildings were valued at well over $400,000. The mansion and buildings alone had a replacement value of over $180,000.

Frederick likely saw the value of having a park and roadways that could bring more people to the other housing developments he was building in the area, since he also donated a two-and-a-half-mile strip of land through the estate for a portion of the Southern State Parkway. Drivers along that stretch of parkway typically take note of the towering pine trees in the median there. These once lined the drive into the original Belmont estate.

Today, there are few remnants from the original Belmont estate; most of the outbuildings were demolished to make way for recreation, playing fields and picnicking. However, walking around the grounds, you might come across the 1879 tombstone of the family's faithful dog Robin near the current park headquarters building.

There are also two cannons that Mrs. August Belmont (née Perry) acquired that are still on the site. They date back to the Battle of Lake Erie, where her uncle led a charge against the British in 1813. Oliver Hazard Perry is famously quoted from his reporting on that battle, when he told then-president James Madison: "We have met the enemy, and they are ours." The cannons are from one of the British ships captured by Perry and his men.

Park Commissioner Robert Moses chose this particular park to establish his regional headquarters on Long Island. It gave him the perfect vista in the center of the island to develop his ambitious plans for the Long Island State Park system. The state originally used the Belmont family mansion as park headquarters. However, in 1935, it razed the mansion and built a new park headquarters. That headquarters building, which contained Robert Moses's former office, is still used for park permit sales, regional park administration and headquarters for the New York State Park Police.

Belmont Lake State Park provides opportunities for picnicking, hiking and biking. There are playgrounds, game courts and fields. Visitors can rent pedal boats, rowboats and kayaks at the large freshwater lake, for which the park is named. The lake also provides anglers the opportunity for springtime trout fishing and summertime bass fishing.

3

BETHPAGE STATE PARK

I n the late 1600s, a Welshman named Thomas Powell bought the land that would encompass Bethpage State Park and its world-famous Bethpage Black Golf Course from Native Massapequas, Secatogues and Matinecocks, who lived in the area. Powell paid £140 sterling for the swath of land that came to be known as the Bethpage Purchase.

The purchase included the land between the towns of Jericho and Jerusalem (now called Wantagh). Many of Thomas Powell's descendants lived on and worked the land, while incoming settlers and their families purchased various other parts of the holdings. Thomas Powell's farmstead spanned from around Powell Avenue to Cherry Avenue and to what would much later become the Grumman Aerospace property. The circa-1690s house originally stood on Hempstead Turnpike, near Route 135 in Bethpage. After Powell's death, his estate passed to his son John Powell and then to a great-nephew, Thomas U. Powell, in 1860. In the decade that followed, the estate and holdings were reduced to a section that is currently within the state park boundaries, although no remnants exist.

There are, however, a few historic Powell family houses still standing in the area. One built by Thomas Powell II in 1700 on what is now Merritt's Road in Bethpage still stands. It is a private home with a historic marker on its grounds. The other is a circa-1850s Powell farmhouse located on the grounds of the nearby Old Bethpage Village Restoration, where visitors can step back in time and see an entire "historic village" composed of buildings from long ago.

Thomas Powell II house. *Farmingdale Public Library and Farmingdale-Bethpage Historical Society.*

In the early 1900s, Benjamin Yoakum purchased the land that later became Bethpage State Park. He acquired 1,368 acres of woodland and farmland, where he built a country estate called Tywacana. The name was loosely based on his birthplace in Texas, called Tehuacana after the Native Tawakoni who once lived there. Yoakum's Tywacana Farms Poultry Company on Long Island sold hatchling eggs and chicks and raised prize cattle and pigs. However, Yoakum was far more than just a chicken farmer.

Yoakum got his start in a railroad surveying gang as a chain bearer for the International–Great Northern Railroad. He later worked with railroad magnate Jay Gould, serving as a land boomer and immigration liaison, bringing Europeans in to settle and work the land in the Southwest. He became a traffic manager and general manager at various railroad lines, including the Gulf, Colorado and Santa Fe lines. Yoakum went on to become general manager of the Frisco line, which included the St. Louis and San Francisco lines. There, he grew the tracks from 1,200 to 6,000 miles before joining the Frisco and Rock Island lines. He created the largest railroad system under singular control at the time. It became known as the Yoakum line.

Yoakum is credited with opening large swaths of the southwestern United States for commerce. There is a town in Texas named after him, and he is quite well known and revered among the railroad enthusiast community.

When Yoakum's later interests turned to farming, he built Tywacana Farms Poultry Company in Bethpage into one of the finest examples of modern poultry farming. The site once held a giant hot water incubator capable of holding thirty-six thousand eggs and a brooder house containing over ten thousand chicks. Free-range chickens roamed on over one hundred acres of land. He strove to raise the finest-quality chick and hatchling stock. The farm was adjacent to the Long Island Railroad tracks for easy transport. Yoakum was an early champion of the "agricultural cooperative movement," which advocated growing and marketing farm products locally to minimize the separation between fresh produce and consumers. He was well known and respected for his Long Island operation.

In 1923, Yoakum hired well-known golf course designer Devereux Emmet to build an eighteen-hole, semiprivate golf course at his Long Island estate, which he leased to Lenox Hills Country Club to run. He also subdivided part of the estate adjacent to the golf course and sold it for residential use.

When Yoakum died in 1929, the Long Island Park Commission expressed interest in acquiring the land for a state park golf course. In 1932, Yoakum's heirs agreed to a sale price of $1.1 million for the 1,368 acres. But the country was in the midst of the Great Depression, and state funding was scarce, so a public benefit corporation called the Bethpage Park Authority was established by the Town of Oyster Bay, Suffolk County and New York State to acquire the property.

Benjamin Yoakum estate. *Farmingdale Public Library and Farmingdale-Bethpage Historical Society.*

In 1934, the federal Works Progress Administration (WPA) put 1,800 men to work to build a clubhouse and three additional golf courses on the site. Joseph H. Burbeck and legendary golf course designer A.W. Tillinghast supervised the work on the four courses, including the extremely difficult Bethpage Black course. Bethpage State Park opened in 1936.

Burback and Tillinghast transformed Yoakum's original Lenox Hills golf course into the Green Course. The clubhouse that once stood there was used to house workers on the new courses until it burned down in 1945. The Yellow Course, designed by Alfred H. Tull, was completed later in 1958.

Today, the park includes five golf courses (Red, Blue, Yellow, Green and Black). Of all the courses, the Black Course is the most heralded. Notoriously difficult and unnerving, it is a masterpiece of golf course design and is ranked among the country's most challenging courses. What sets it apart from other courses is that it has the distinction of being the only publicly accessible course ever used for the prestigious U.S. Open Championship game, held in 2002. At that time, well-known golf course architect Rees Jones prepared the course for the championship. The original 1930s clubhouse, built in the WPA era, was refurbished and updated, and new benches and signs were added throughout the property. New trees and sod were added to the course, and the Pro Shop was expanded to become one of the most comprehensive shops in the country.

During the legendary 102nd U.S. Open held at the course in 2002, Tiger Woods captured his second U.S. Open title and eighth major championship win over Phil Mickelson by three strokes. It was one of the most heavily attended U.S. Open Championships ever. Bethpage Black went on to host the 2009 U.S. Open, in addition to the 2012 and 2016 Barclays and the PGA Championship in 2019.

Golfers love and revile the Black Course for many reasons. It is a long course at 7,459 yards, and it has only one par-3 under 200 yards. It is hilly, its "rough" is quite rough and the bunkers are broad and deep. To play the course is a workout, to state it mildly. Today, the public can get a tee time if they think they are up for the challenge. A round costs approximately forty dollars. However, a sign at the course reads: "Warning: The Black Course Is An Extremely Difficult Course Which We Recommend Only For Highly Skilled Golfers."

In addition to the golf courses at the park, today, there are also public picnic facilities, playing fields, tennis courts, bridle paths, hiking and biking trails and cross-county skiing trails. Catering is available in the circa-1930s clubhouse, and a driving range is available for public use.

4

BRENTWOOD STATE PARK

Beneath the synthetic turf fields where teams at Brentwood State Park play competitive sports, there was once a farm that supplied around two thousand tons of produce per year to feed residents of the nearby Pilgrim State Psychiatric Hospital.

The farm was located between Crooked Hill Road and Wick's Road and encompassed the current Brentwood State Park and part of Suffolk County Community College, located adjacent to the park. It included a piggery, horse barn, several staff cottages and produce storage buildings. Evidence of the farm can still be found on the college campus. Two of the farm cottages still stand and serve as storage and maintenance facilities for campus use. The college also uses one of the other family cottages, now called South Cottage, as a study hall for honor students. Two other buildings from the former hospital farm, each with distinctive, forward-pointing gables, still stand on the campus. Farmworkers used these buildings as a barn and hayloft, as evidenced by a hay-lifting mechanism still visible in the soffit and obvious patchwork on what appears to be the loft. A New York State Office of Mental Health (OMH) map shows another building on the adjacent college ground, labeled "Farm Colony Building." Still standing, it is constructed of brick, in much the same style as the other buildings on the main Pilgrim State Hospital site. It has ornate arches over its exterior doors and unique concrete sculptures of farm animals built into its architecture, giving a clue to its former use. It has had various other uses over the years, including patient housing. There was also once a morgue in the basement.

Pilgrim State farm and farm buildings (*upper center*). *Queens Borough Public Library, Long Island Division, Long Island Daily Press Photograph, Morgue Collection.*

Today, the college calls it Caumsett Hall, and it is used mostly for general administrative purposes.

The OMH map of the farm area also shows there was a later geriatric care facility for men and two continued care treatment buildings for women in the area of the state park playing fields. These have been demolished.

Pilgrim State opened in 1931, decades after the establishment of other rural farm colonies at Kings Park and Central Islip, which were built in the late 1800s. The farm colonies were a great public experiment, as they were meant to bring mentally ill patients into a country setting where they could productively work the land and contribute to the functioning of the facilities. The manual labor and fresh air were part of their treatment and rehabilitation. Patients were trained to perform certain tasks and were taught skills that would be useful upon their release.

Previously, New York grouped the mentally ill with common criminals and packed them into "poorhouses" or "lunatic asylums," where they were often confined under harsh conditions. However, near the turn of the twentieth century, new schools of thought began to emerge that recognized mental illness as a disease of the brain, just as diseases attack other organs of the body. Thomas S. Kirkbride is recognized as one of the greatest early advocates for more humane treatment of the mentally

Farm cottages now at Suffolk County Community College–Brentwood, adjacent to the park. *Author's collection.*

ill in New York. His Quaker upbringing in Pennsylvania and postgraduate experience working in a Quaker hospital for the insane taught him that that kindness and strict routine could be used in the treatment of mental illness instead of punishment and confinement. His learnings were based on the studies of two doctors who worked independently from each other, William Tuke and Philippe Pinel. Both chose to unchain patients in an attempt to treat them with moral kindness and structure. The term *asylum* was changed to *hospital* to reflect the growing understanding that mental illness was an illness that should receive medical treatment. Kirkbride took this understanding and went so far as to develop a plan outlining how hospitals for the insane should be constructed. It described how windows should open to prevailing winds and that the hospitals should be situated in pastoral settings with good land for tilling and access to water. In New York, land was cheaper out in the country, and it was believed that Long Island, still quite rural and agrarian at the turn of the century, would be healthier for the treatment of those in need.

New York State built Pilgrim State Psychiatric Hospital specifically to ease the burden at Kings Park and Central Islip Hospitals, as the wards of New York City continued to spill outward to Long Island. In the year it opened, Pilgrim took in over two thousand patients. Over the next twenty years, its patient population grew to over thirteen thousand. At its peak, Pilgrim was the largest psychiatric facility of its kind. It was named for Dr. Charles Winfield Pilgrim, who spent fifty years of his life caring for the mentally ill.

The campus sprawled from Long Island Avenue, between Commack Road and Wick's Road, nearly up to the Long Island Expressway. While it was vastly larger than the other farm colonies, Pilgrim was designed to be as self-sufficient as possible, generating its own electricity, pumping its own water and growing its own food. The hospital community had its own heating plant, sewage system, fire department, police department, courts, church, post office, cemetery, laundry, store, amusement hall, athletic fields and greenhouses.

Pilgrim's patients were generally separated by gender and affliction. Some chronic and physically disabled patients were housed in continual care facilities, while others were allowed freedom to participate in communal activities, including sports and exercise, games, theater, dancing, music, arts and crafts, woodworking and other activities. While the hope was that many of the patients could be cured and released, it was understood that many of them would likely require care for the rest of their lives. Despite some of the more horrific perceptions of psychiatric facility confinement, there were efforts made by the administrators and staff at Pilgrim to create a pleasant atmosphere to the extent possible on limited budgets. There are many historic images of staff putting up holiday decorations and holding holiday pageants, dinners, plays and dances and the patients participating in shows and entertainment.

Occupational therapy was an important component of the treatment at Pilgrim. Many patients were assigned to learn usable skills based on their physical and mental capabilities. This could include everything from basket weaving, embroidery, shoemaking, mattress making and clothes making to woodworking, cabinetry, equipment repair and land clearing. Food was prepared on site, and patients often worked alongside the cooks. There was even a print shop where patients worked publishing newspapers and bulletins for patients and staff. There were plenty of outdoor tasks, including tilling, planting, weeding and harvesting if patients chose to work at what is now Brentwood State Park and the Suffolk County Community College campus.

The psychiatric hospitals on Long Island abandoned the working "farm colony" concept when New York State ruled that they could not put patients to work without pay. Afterward, patients received care in a more "institutionalized" setting. By the middle of the twentieth century, New York State began discharging many former patients into the community as newer forms of treatments and medications emerged, and the need for large facilities to treat the mentally ill rapidly declined. In 1996, the state consolidated operations and relocated patients from nearby Kings Park Psychiatric Center and Central Islip Psychiatric Center to what remained of the Pilgrim campus. Although Pilgrim is greatly downsized, a portion of it remains open, with 278 inpatient beds and many outpatient psychiatric, residential and related services.

Today, the Pilgrim campus and the surrounding area is an odd juxtaposition of abandoned buildings, decaying architecture and the rubble of former buildings that a proposed housing developer unceremoniously tore down and left in piles on the grounds. Brentwood State Park and the state-owned Suffolk County Community College made use of some of the vast acreage. All around are clues to the size and scope of the facility, including a former power plant, a water tower, numerous abandoned buildings and a train station and spur now hidden in the woods.

Much of the oak brush woods adjacent to the former psychiatric facility have been preserved, notably around Commack Road and the former thirteen-story Edgewood/Mason General building that was demolished in 1990. That building once held patients who were returning from the trauma of World War II. Now called Edgewood Oak Brush Plains State Forest, the park comprises 813 acres of woods and fields with over 23 miles of trails. The preserve is composed of rare pitch pine/scrub oak barrens that contain dense shrub thickets. There are only three pine barren regions on Long Island, and they are recognized as globally rare occurrences.

Brentwood State Park initially opened to the public in 2003, and the athletic fields, including baseball and soccer fields, concession stands, restrooms and playgrounds were completed in 2013. Today, the park is a state-of-the-art sports campus with eight synthetic turf soccer fields to accommodate all division age groups, including two adult, one intermediate, three bantam and two junior fields. The two synthetic turf baseball fields are regulation size, with ninety-foot diamonds built for high school players and older players. There is also professional field lighting for nighttime play.

While Brentwood State Park is smaller in acreage than most other Long Island State Parks—and clearly has a predominantly sports-oriented usage—the acquisition of the land for the park is part of New York State's ongoing commitment to acquiring parcels for parkland on an opportunistic basis. In this case, the New York State Office of Mental Health transferred the land for Brentwood State Park to New York State Office of Parks, Recreation and Historic Preservation for the public's use.

5

BROOKHAVEN STATE PARK

Brookhaven State Park was once part of a vast encampment and military training complex called Camp Upton that the government created to urgently equip and prepare large numbers of men for battle as America's involvement in World War I became imminent.

Prior to the establishment of Camp Upton, the military forces of the United States were primarily composed of local militias, state-regulated regiments and regional armies that were responsible for managing domestic engagements. The most notable among these was the Continental army, which was formed during the Revolutionary War and disbanded after the war was over. There were occasional attempts to create a fully organized and operational standing national military during peacetime. However, the Great War in Europe propelled a massive mobilization effort to create the United States' first national army that would be large enough to handle the upcoming combat in Europe.

Camp Upton became one of sixteen cantonments established across the country. It was named for Major General Emery Upton, who distinguished himself in the U.S. Civil War commanding artillery, cavalry and infantry units. He was a strong proponent of America's need to establish a regular, organized, nationwide army with regular staffing, recruitment and military education.

Laborers and army recruits swiftly built the camp on Long Island between August and December 1917. It was commanded by Major General J. Franklin Bell upon completion. The garrison unit was the 152nd Depot Brigade, which accepted nearly forty thousand recruits from New York, New

Recruits being mustered at Camp Upton. *Longwood Public Library*.

Jersey and Connecticut to prepare them for service. A sixteen-week training period prepared men for trench warfare, grenades, tanks, machine guns and hand-to-hand combat before they were shipped overseas for battle.

Many of the recruits became part of World War I's Seventy-Seventh Infantry Division, widely known for their valor in the woods of Argonne, France, where the so-called Lost Battalion fought heroically despite overwhelming obstacles.

The Seventy-Seventh Infantry was composed largely of drafted recruits from New York City. They were neither army nor national guardsmen, but they were instead known as a "national army," also nicknamed the "Times Square Division," "Liberty Division" and "Metropolitan Division." These and other American soldiers became part of the largest American offensive in history in September 1918, which involved over 1.2 million soldiers in France's Meuse-Argonne region.

The story of the 77th Infantry's Lost Battalion became legendary. The soldiers were from a number of companies led by Major Charles Whittlesey alongside the Second Battalion of the 308th, led by Captain George McMurtry. McMurtry fought previously in the Battle of San Juan Hill with President Theodore Roosevelt's Rough Riders during the Spanish-American War.

As the story goes, the troops became separated from other nearby units in the Argonne woods as hostile German forces surrounded them on all sides. All lines of communication were lost, and while coming under heavy fire, they used a carrier pigeon named Cher Ami to send word to command requesting assistance. Cher Ami delivered the message, and the Americans regrouped to beat back the enemy. The remaining 190 men, all heroes, were eventually able to walk out, although hundreds of their unit were shot and killed. The newspapers called them the Lost Battalion, and Americans at home raptly followed and knew their story.

One of the Seventy-Seventh Infantry Division's most famous recruits was Irving Berlin, the composer known best for his song "How I Hate to Get Up in the Morning," which he wrote as part of his Broadway play about his experiences at Camp Upton called *Yip, Yip, Yaphank*.

After World War I, the property briefly served as a demobilization site for returning veterans and as a recruit educational center to educate illiterate and "alien" enlistees, primarily of Eastern and Southern European descent. During World War I, the military recruited many of these new immigrants to the war cause, even though they did not speak much English. After the war, the army school at Camp Upton served to "Americanize" new enlistees by training them in citizenship and offering those who committed to three years of service the opportunity to become naturalized citizens. They were taught English and were drilled in self-discipline, military protocol, United States history, politics and geography. They were also taught about the benefits of capitalism to ease their transition into the American way of life.

Field bakery at Camp Upton. *Longwood Public Library.*

By 1925, the military no longer had any use for the nearly six thousand acres of land. At the time, the property was considered largely a barren waste, since much of the land had been violated and practically no topsoil remained. The New York State Conservation Commission reforested the land over the course of twenty years using techniques that the French had devised to reforest their degraded areas after the war. This included the planting of pine trees, including Austrian pine and maritime pine, which were specially adapted to grow in sandy soil. The long-term plan was to turn the land into a large game sanctuary.

In 1940, it became apparent that another training facility might be necessary as World War II loomed. Camp Upton served as a draft center at that time to train inductees for before operations were eventually shifted to Fort Dix in New Jersey.

Also in the early 1940s, Camp Upton may have been used as an internment camp to hold Japanese residents from New York when the war broke out. It is believed it was called the United States Defense Aliens Japanese Internment Camp Upton in New York, although some lists of internment camps in the United States show Camp Upton as holding one thousand persons of unknown national origins, which may have included Germans and Italians.

In 1944, while World War II was still raging, the camp became a convalescent hospital and rehabilitation center for wounded veterans. It provided exceptional care for the time, as heavily wounded and shell-shocked men returned to the United States. In addition to medical care, the hospital offered the men recreational facilities, including swimming pools and tennis courts, to occupy them as they recovered.

In 1947, the United States government selected Camp Upton to become the site of a large-scale research center for the peaceful use of atomic energy and other emerging technologies. Brookhaven National Laboratories was built on the southern portion of the property in coordination with Massachusetts Institute of Technology (MIT) and other East Coast Ivy League colleges to become a large-scale, state-of-the-art research facility capable of attracting the best scientific and industrial minds from around the world. Initially, the former military barracks buildings were used to house scientific equipment. After the war, Los Alamos scientists and researchers came to Brookhaven National Lab, where they applied technology gained in the production of the atomic bomb toward peacetime production of medical isotopes and other scientific, industrial and agricultural materials until 1963.

Today, Brookhaven National Lab sits on the former Camp Upton site south of Middle Country Road. Thousands of residents and visiting researchers explore applications related to chemistry, biology, physics, the environment and energy at the facility. Brookhaven National Lab has produced over a half-dozen Nobel Prizes winners and has a dozen other Nobel Prize connections.

In 1971, the federal government agreed to hand over 1,638 acres on the north side to the Long Island State Park Commission to create Brookhaven State Park. Much of the property is untouched, containing trails for hiking, biking and seasonal hunting. The Oak Pine Forest of Brookhaven State Park is part of the larger Long Island Pine Barrens region. Surrounding it are other pine barren preserves, including the Rocky Point Pine Barrens State Forest, the Robert Cushman Murphy County Park and the Long Island State Pine Barrens Preserve. It also contains native dogwood, wild geranium, fern and pitch pine, along with wetlands and coastal plain ponds. Overall, there are five marked trails ranging in length from three miles to five miles in the park, along with another sixteen miles of unmarked trails. Caution is advised, as the trails are primarily composed of sand and compact sand, which is a challenging terrain for long hikes and bikes.

CALEB SMITH STATE PARK PRESERVE

Long before Caleb Smith State Park became a park, the affluent members of the Wyandanch Club constructed an imposing mansion on land that once belonged to an early Smithtown settler to use as their own private sportsmen's retreat. They utilized the hundreds of acres of surrounding woodlands and streams for hunting, fishing and unwinding.

During the early twentieth century, Long Island was still quite rural, and there were several hunting clubs in the area. Wealthy bankers, tycoons, industrialists and others formed these clubs to escape the confines of their wealth and obligations in the city and to act as sportsmen and enjoy the comradery of like-minded companions.

The Wyandanch Club was originally called the Brooklyn Gun Club, and its membership comprised wealthy businessmen from Brooklyn, New York, who had organized themselves around monthly pigeon shoots in Dexter Park, which used to be located on the west end of Woodhaven, New York, on Jamaica Avenue. By the late 1880s, the Brooklyn Gun Club, which was limited to forty members, had become exceedingly wealthier and decided it would seek a rural setting for its hunting game activities.

The homestead the club purchased in Smithtown was then called the Theodorus Smith Place at Willow Pond. Different generations of Smiths lived in and expanded on the original house, built in the pre–Revolutionary War era around 1752 for Caleb Smith as a gift from his father, Daniel. Throughout the mid-1700s, Caleb Smith served as a judge for the Court of Common Pleas of Suffolk County, a Suffolk County supervisor and a

The original Caleb Smith house, with two rooftop window dormers, can be seen in the center, surrounded by the later Wyandanch Club additions. *Author's collection.*

member of the New York State Assembly. He was also a Revolution-era Patriot and was the great-grandson of Richard "Bull" Smith, the founder of the Town of Smithtown.

Local Smithtown residents revere the legend of Caleb Smith's great-grandfather Richard "Bull" Smith. Local legend says that a Native Montaukett sachem called Wyandanch presented the land to Lion Gardiner, an early settler who helped rescue Wyandanch's kidnapped daughter. Richard Smith, or Smythe, acquired the land from Gardiner in 1663, at which time, he rode his bull Whisper around the entire perimeter of what would become Smithtown to mark its boundaries. He rode from sunup to sundown, stopping to have some bread and cheese along what would become the westernmost boundary of Smithtown. The road there is still called Bread and Cheese Hollow Road.

To commemorate the great riding of the bull, Lawrence Smith Butler, a descendant of the town's founder, commissioned a giant cast-bronze statue of Whisper the bull. The Town of Smithtown held a dedication ceremony on May 10, 1941, on the plot of land between the fork of Route 25 and Route 25a, just east of Caleb Smith State Park, where the statue of Whisper the bull still stands.

The Boss Long Island Fisherman at the Wyandanch Club. Smithtown, L. I.

No. 1276   Published by The Long Island News Company, Flushing, L. I.-Leipzig-Berlin

A member of the Wyandanch Club. *Richard H. Handley Collection of Long Island memorabilia, Smithtown Library.*

When the Brooklyn Gun Club bought the house and seven hundred acres from Caleb Smith's descendants back in 1888, it was still in excellent shape, with sturdy, hand-hewn wood beams and solid timber construction. The club decided there was no reason to tear it down or rebuild it. Instead, it built its club around it. If you look closely at the front of the clubhouse, you can see the original Smith farmhouse in the center of the building, just offset to the left. There is also a second original structure located just behind the main house that members incorporated into the much larger clubhouse additions.

By 1895, the club had fitted the structure with modern conveniences and greatly expanded its size. At that time, the clubhouse had a billiard room, kitchen facilities, public areas, a porch, a reading room, a large central fireplace and accommodations for members. They stocked the property with quail brought in from Tennessee and filled the ponds and streams with trout spawned in an on-site hatchery on the property. At one point, the club leased another eleven thousand acres from farmers in the area to broaden their sporting activities.

They adorned the clubhouse with guns, fishing poles and taxidermy exhibits. Members could look at the kills and recall and relay their stories to others. The house's dark wood paneling and molding throughout creates a warm and rustic lodge feel. Today, if you visit the clubhouse, you can see the restored lockers, where members of the club would store their individual stashes of booze for use when they came out to relax.

The club also saw fit to preserve an important part of Smithtown's local history inside the clubhouse. A large and heavy aged oak door stands out from the rest of the house's rustic décor. It originally served as the exterior door of the historic 1700s farmhouse. During the Revolutionary War, when Long Island was occupied by the British, a British soldier sought entry to the house, and according to legend, the door was promptly slammed in his face and barred shut. The British soldier thrust his bayonet at the door in frustration. The door, with its damage, is still visible.

When the Brooklyn Gun Club chose to change its name in the 1890s, it selected the name Wyandanch Club as a tribute to Chief Wyandanch, whom it recognized as being inextricably linked with the earliest history of the area. Inside the mansion today, you can still see an impressive woodcarving of Chief Wyandanch's likeness that the club created and placed over the fireplace in tribute.

An impressive woodcarving of Chief Wyandanch's likeness over the fireplace at the Wyandanch Club. *Author's collection.*

New York State bought the property from the Wyandanch Club in 1963, as Long Island was becoming increasingly suburban in character. While the park was originally named Wyandanch Preserve and then Nissequogue River State Park, the state later changed its name to Caleb Smith State Park and Preserve to honor its historic connection to the Smithtown area. The park opened to the public in 1974.

Today, the 270-plus-year-old house/Wyandanch Club mansion still stands and is used as an educational center that contains the history of the original Smith house, including its time as a hunting club, and the history of the flora and fauna of the area, along with maps, dioramas and natural history/taxidermy displays. The park offers nature programs for adults as well as children.

The grounds include 543 acres of undeveloped landscape and many different habitats, including freshwater wetlands, ponds, streams, fields and upland woods crisscrossed with hike and bike trails. The park is also popular with birdwatchers. Fly-fishing is permitted with a license on the Nissequogue River during certain times of the year. There are no playing fields or courts for recreation; rather, the park invites visitors to enjoy the quiet solitude of nature that members of the Wyandanch Club must have appreciated in centuries past.

1
CAMP HERO STATE PARK

Camp Hero State Park is located in Montauk, a once-small fishing village on the easternmost point of Long Island famous for its stunning white sand beaches that attract throngs of tourists during the summer season.

However, Montauk has a centuries-long, larger-than-life history as a center of military activity. Within the boundaries of Camp Hero State Park, remnants of this military past are still on display. A large, decommissioned radar tower and concrete bunkers serve as a reminder of its past military use. Montauk's military history dates back even further, to the Revolutionary War era. At that time, British troops that occupied Long Island recognized its strategic location as an ideal launch spot for the Siege of Boston in 1775, as well as during their attempts to create a blockade off the Connecticut coast. Additionally, after the Spanish-American War in 1898, Montauk's remoteness made it a logical choice for Colonel Theodore Roosevelt and his Rough Riders of the First U.S. Volunteer Infantry to quarantine around nearby Fort Pond upon their return from battle.

Montauk's location at the eastern tip of Long Island also made it an ideal outpost for surveillance. During World War I, the government created a naval air station to send reconnaissance aircraft, including dirigibles and planes, to guard New York waters. Previously, the navy had no air patrol capability except for one aeronautic station in Pensacola, Florida, which was not nearly large enough to train aviators for monitoring the expansive and potentially vulnerable East Coast and New York metro area. It was Rear Admiral Nathaniel Usher who envisioned a series of ten aeronautic stations

Sergeant Fred Houseknecht at Camp Hero with a dog and sixteen-inch casement gun, 1941 (World War II). *Alice Houseknecht Collection, Montauk Library Archives.*

from Ocean Beach, New Jersey, to New Haven, Connecticut, and out to Montauk Point. Montauk was a top priority, and the admiral recommended a base of seventy-two airplanes and two dirigibles. By 1918, Montauk had a functional aerial patrol on about thirty acres, along with a reconnaissance station to house coast guard personnel and troops.

During World War II, the possibility of an East Coast invasion seemed even more threatening. Along the shipping lanes off Montauk and up through New England, Nazi U-boats patrolled the waters, sinking cargo ships and fuel tankers. By 1942, German submarines had destroyed over 170 ships along the coast from the North Atlantic to Florida, including the *Norness*, the first tanker sunk in New York waters, sixty miles off Montauk Point, Long Island.

Nazis had even begun to infiltrate Long Island. On June 23, 1942, a group of German saboteurs landed in Amagansett, near Montauk, with explosives to destroy power plants and bridges. One member of the group, whose mission was called Operation Pastorius, surrendered to the authorities, and the others were arrested and subsequently executed. The need to protect Long Island and New York City became urgent. The government recommissioned Camp Hero in 1942 and outfitted it with heavy artillery and machine guns to defend the coast against possible German submarine attacks.

In August 1942, Battery 113, also called Battery Dunn, named for Colonel John M. Dunn, was installed at the site. The remnants of this battery are still visible at Camp Hero. It, along with Battery 112, contained two sixteen-inch guns surrounded by a six-hundred-foot-long bunker made of reinforced concrete to protect against aerial or naval attack. Another, Battery 216, had two six-inch guns for added protection.

The bunker was entirely self-sufficient, with its own electricity, ventilation system and water supply. Named Camp Hero after Major General Andrew Hero Jr., who served as chief of coastal artillery from 1926 to 1930, the site also served as a large torpedo testing facility.

Cleverly, the entire self-sustained camp was camouflaged to resemble a New England fishing village to escape notice from potential enemies prowling the Eastern Seaboard. The artillery was hidden by foliage, and the concrete bunkers had a wooden-like façade and fake windows painted on it for disguise. A recreational hall was designed to look like a New England church with a steeple. Remnants of the gymnasium/church are still present at the site.

Six hundred men and thirty-two officers lived in the "seaside village" through the end of the war in 1945. By 1947, the U.S. Army had deactivated the site and removed the heavy artillery.

Used briefly as a training facility by the army reserves, the camp was largely abandoned until the 1950s, when a portion of it was used for surveillance during the Cold War. The 773rd Aircraft Control and Warning (AC&W) Squadron (later the 26th Air Division/Montauk Air Force Station) was stationed there, and radar equipment was installed to monitor and identify aircraft in the area.

From 1951 to 1957, it served as an antiaircraft artillery training site for the U.S. Air Force units from the New York area during the Cold War. The training included firing 120-millimeter antiaircraft guns at radio-controlled "planes" flown over the Atlantic Ocean and at targets sent out on barges up to twenty thousand yards away.

There were groundbreaking technologies tested and used at the site. This apparently included wide area networks (WAN), magnetic memory storage, keyboards and modular circuitry containment. The SAGE (semiautomatic ground environment) radar system was installed in 1958, including the AN/FPS-35 radar antenna, built by Sperry. SAGE was the country's first air defense system capable of condensing data from multiple

AN/FPS-35 radar antenna, Montauk Air Force Station, 1980. *Ed Crasky Montauk Air Force Station Photographs, Montauk Library Archives.*

radar sites. The antenna had a 200-mile range and was so powerful it used to disrupt local TV and radio broadcasts. The 120-foot-wide antenna is still visible at the site.

Security was quite intense around the facility. That, combined with the television and radio disruptions by the antenna, likely fueled speculation and later conspiracy theories about the site. Some claim the government experimented with time travel, mind-control and teleportation, and some say that there were even aliens housed at the facility.

The facility was closed in 1982 as newer radar technology became available. The site was largely vacant, and at one point, the land was almost sold to a developer. Local opposition, combined with potentially huge cleanup costs, dampened any thoughts of massive redevelopment. Between 1974 and 1984, the government transitioned the property from military use to use by local, state and federal agencies. In 1992, 412 acres were designated as public access state parkland, and in 2002, Camp Hero State Park opened to the public.

Before the park opened to the public, remedial work was conducted that included the removal of some buildings, underground storage facilities, equipment and electronics. The U.S. Army Corp of Engineers conducted extensive safety and contamination testing on the site. The massive bunkers that still stand are sealed up with solid concrete, so the "Do Not Enter" signs seem to state the obvious futility of trying. However, visitors are welcome to walk around the passive-use site and read the history on historical signage throughout.

Today, where the military once held exercises to defend the country and prepare it for national wartime engagement, the landscape now comprises maritime forests, freshwater wetlands and stunning views of the Atlantic Ocean from dramatic bluffs towering above the beach below.

The site is now highly regarded for its surf fishing, and year-round, there are people fishing at the ocean's edge for striped bass and bluefish by permit only. The shoreline is treacherous, and there are no lifeguards on duty, so swimming is forbidden. Instead, visitors can enjoy an extensive system of trails for hiking and biking and can explore a site that holds a deep and respected place in history for defending the security of the United States coastline.

8

CAPTREE STATE PARK

Off the South Shore of Long Island, there are a series of mostly undeveloped barrier islands running from the East Rockaway Inlet at Long Beach to the Shinnecock Inlet near Hampton Bays. They protect Long Island from the never-ending, pounding surf of the Atlantic Ocean.

While remote and accessible only via boat or bridge, there has long been activity here dating to Natives and early settlers who recognized the area's opportunities for whaling, fishing and shellfish harvesting. Beginning in the late 1800s and early 1900s, more people began venturing to these barrier islands to enjoy the ocean surf, boating and fishing opportunities. Small communities sprang up over the years, bringing with them modest hotels, marinas and other economic pursuits.

Captree State Park is situated at the easternmost point of the barrier island that stretches all the way from Jones Beach in the west. At one time, Captree was home to a handful of notable establishments that no longer exist. These include a sportsmen's club called WaWa Yanda that dated to 1878. WaWa Yanda, which is a play on "way, way yonder," attracted politicians, investment bankers and industry tycoons from New York City and the surrounding area. These included President Chester A. Arthur, Mayor Robert Van Wyck and Governor Alfred E. Smith, among others. The club started as a retreat for sporting activities and grew into a fabulous resort for the wealthy. It was known for its lavish yet informal parties that often included seafood feasts, clambakes, fish fries and abundant alcohol

Image of the WaWa Yanda Club exterior, Captree Island, circa 1950. *Town of Babylon, Office of Historic Services.*

consumption, according to reports at the time. Invitations to visit the club were quite coveted. WaWa Yanda remained popular until the 1920s. When Prohibition banned the sale of alcoholic beverages, the site may have served as a rumrunning stop, although business had dropped off substantially by that time. In the 1950s, a terrible storm undermined the building and set it into decline. It was torn down in the 1970s.

Just south of WaWa Yanda, Henry Havemeyer, the son of New York City mayor William Havemeyer, had a private resort located on the east side of Captree State Park. Havemeyer made his money as a partner in Havemeyer and Vegelin Wholesale Tobacco Company and as a trustee of the Havemeyer Sugar Refining Company in Williamsburg, Brooklyn. He also served as president of the Long Island Rail Road.

His summer resort was known as the Armory, and it was host to an invitation-only crowd of friends and business associates in the 1880s. The Armory was located close to what is now the New York State Boat Channel near the marina at Captree State Park. It lasted until Havemeyer died in 1886. Nearby was Conklin's, which burned down in the 1880s.

East of the upper bathhouse at Captree State Park, near the current fishing pier, a large-scale fishery once stood, enriched by the bounty of the bay and ocean. Today, that operation no longer exists, but many individual anglers still cast their lines from the docks in the same spot.

In 1902, the U.S. Army Corp of Engineers constructed a boat channel that bisected Captree Island. Today, visitors to Captree State Park cross over a small drawbridge where the land was once naturally connected. On the north side of the channel, a small community developed in the 1890s, when the Oak Island Beach Association sought to create a resort and community there. A plan for a religious retreat got off to a start, and an auditorium and small hotel called Oak Island Hotel were built. There were later plans to create a resort and casino, but by the 1920s, erosion and other economic challenges had led to the demise of any large-scale development ideas. Today, a small community of cottages on Captree Island is all that remains, along with another community of summer cottages on the south side at Oak Beach.

In the community of Oak Beach, just west of Captree State Park, you can see a remnant of a time when this area was patrolled by brave men who risked their lives to save victims of the numerous shipwrecks in that area. This part of the barrier island was home to U.S. Life-Saving Service (USLSS) Station no. 26 Oak Island Beach/Coast Guard Station no. 84, built in 1872. Station Oak Island Beach conducted general patrol duties along what is now Captree State and participated in many daring watercraft rescue operations. Out of the twenty-three USLSS stations built along the South Shore of Long Island in the 1870s, the Oak Island Beach Station is one of the only original red house–style stations still in existence. Stylistically, it was called a red house because of its red roof, which was standard for USLSS stations.

The Oak Island Beach Station House was closed by 1937, with the last serving keeper listed as Chief Boatswains Mate W.J. Eldridge. The station was temporarily reactivated during World War II but was abandoned again, and the General Services Administration transferred the land back to the Town of Babylon in 1954. Today, the station on Oak Beach Road serves as the Oak Beach Community Center and still bears its traditional red roof. Many of its original architectural details are intact, and it contains a small, well-done exhibit detailing the work of the station.

By the time the Long Island State Park Commission acquired the land for Captree State Park in 1930, it had already acquired land for Jones Beach State Park, the Jones Beach Causeway, Gilgo Beach State Park, Ocean Parkway and Fire Island State Park (now Robert Moses State Park), along with Southern State Parkway and Northern State Parkway.

Ferry from Captree State Park to Fire Island before the bridge was built. *Long Island State Parks.*

The commission originally planned to build a bridge from Captree State Park to Fire Island State Park and, from there, construct a scenic ocean parkway through Fire Island to the east end of Long Island. Protestors thwarted those plans. Instead, the commission proposed a scaled-back plan to build a parkway from Southern State Parkway and a two-mile-long bridge across the Great South Bay to Captree State Park. From Captree, regular hourly ferries would transport visitors across the narrow inlet to the fantastic ocean beaches on the Fire Island barrier island in about fifteen minutes. Previously, Fire Island was accessible only by private boat or via ferries that ran intermittently from Babylon or Bayshore and took over an hour. The New York State Post-War Planning Commission approved the revised plan in 1944, although the state would not construct the parkway and bridge for another ten years.

After construction was complete, Captree had 1,200 parking spaces, where visitors could park and board the fast and convenient ferries to the ocean beaches. In 1964, the state completed the final span from Captree State Park to Fire Island, and there was no longer any need for the ferries.

A restaurant offering seafood and other fare still stands at Captree. The marina at the park, located on the State Boat Channel, grew to encompass the largest public fishing fleet on Long Island. Independently owned and operated, the fleet at Captree still offers private charter and open boat saltwater fishing opportunities, along with sightseeing and scuba excursions. There are also opportunities for fishing along the piers for permit holders.

Much of Captree State Park's three hundred acres are undeveloped brushland, and its shorefront is home to many nesting species of birds. Swimming is not permitted at Captree State Park, but there are scenic picnic areas with tables and barbecues overlooking the Fire Island Inlet, along with an accessible playground and boat launch ramp. Kayaking and paddle boarding are permitted in certain areas.

CAUMSETT STATE HISTORIC PARK PRESERVE

When he was just twenty-nine years old, Marshall Field III purchased 1,750 acres of property in Lloyd Harbor and built the large, English-style country manor that still stands today at Caumsett State Park. Field III named the estate Caumsett, meaning "place by a sharp rock," which is what the Native Matinecock called it.

As early as 1654, the Matinecocks sold the land to three English men, Samuel Mayo, Daniel Whitehead and Peter Wright. The land, which juts out into the Long Island Sound on the North Shore, was purchased by James Lloyd in 1684 and was used as farming and grazing land. James's son Henry built a saltbox house on the property in 1711, as did his grandson Joseph in 1766. Both of these early Lloyd houses still stand at the entrance to Caumsett State Park.

During the Revolutionary War, the British occupied the area and had two forts on the neck, Fort Franklin, which was the site of one of the largest American offensives against the British on Long Island, and nearby Target Rock, which is currently a National Wildlife Refuge. Target Rock is named for a large rock that stands just offshore that the British used for target practice.

Marshall Field III purchased the waterfront property at the northern edge of the neck to create his grand estate in 1921. Field III was the grandson of Marshall Field. In the mid-1850s, Marshall Field and Company pioneered the concept of the modern department store, which hit the right note with leisurely and refined upper-class customers. Marshall Field Department

Marshall Field estate aerial view. *Long Island State Parks.*

Store pioneered such amenities as a coat check, women's lounges, tearooms, telephones and restaurants in a cheery, low-pressure sales environment. The company would become one of the largest retail chains in the country, known and beloved in the Chicago-area market. The business was valued at over $30 million by the mid-1880s.

Marshall Field III's grandfather died a very wealthy man in 1906 at the age of seventy-one. His will stipulated that his great fortune be passed to his grandsons, Field III and his brother, Henry, since their father, Marshall Field II, had died tragically the year before of a gunshot wound; he was only thirty-seven. The family said the terrible accident occurred while he was cleaning his gun. The fortune was held in trust for nearly forty years, since the two heirs were both young boys at the time their father and grandfather died. Their mother remarried and sent the boys to England for schooling, although both returned to America to join the family business.

Field III's brother, Henry, died in 1917 at the age of twenty-two after an operation, and the estate in its entirety was entrusted to Marshal Field III. It was not until 1943 that Marshall Field inherited the $75 million from his grandfather's business. Field III had already built a personal fortune with interests in banking, publishing, horseracing, retail and numerous other holdings under his Field Enterprises Inc. He would later go on to found the liberal-minded *Chicago Sun* newspaper (later the *Chicago Sun-Times*). While he was quite wealthy, he was, interestingly, an active supporter of Franklin Delano Roosevelt and his New Deal to increase taxes on the elite to fund federal work projects and other initiatives to benefit Americans during the Great Depression.

Marshall Field III enlisted the services of famous architect John Russell Pope, who designed the Jefferson Memorial and National Archives in Washington, D.C., to build his grand estate on Long Island's North Shore. The area is known as the Gold Coast, due to its large concentration of wealth and vast number of estates. Pope situated the large, Georgian-style brick mansion to overlook a vast expanse of grazing fields and woodlands. The estate included a hunting preserve, a farm, a stable, a greenhouse, outbuildings for staff and its own water and electrical supply.

Field III commissioned renowned landscape design architects the Olmsted brothers to create the gracious greenery on what previously were overgrown woodlands, meadowlands and farmlands. The property had a massive dock where Field III could moor his yacht, a pool and an indoor tennis court. The family sometimes threw lavish parties, including many fundraisers and benefits to aid various societies and causes. Field III and his family did not spend an extensive amount of time at the estate, as he owned several other properties in New York, Chicago and elsewhere. While many of Field III's Gold Coast neighbors lost their great fortunes during the Great Depression, Field III's Caumsett estate survived the era.

Field III was married three times, lastly to Ruth Pruyn Phipps. After Field III passed away in 1956 at the age of sixty-three, his third wife placed the estate up for auction. Interestingly, she accepted a bid from the Long Island State Park Commission in 1960, despite a real estate developer's offer for $1 million more to build a housing development.

The Long Island State Park Commission originally intended to create an active-use park at the site, with a public beach, boardwalk, bathhouse, picnic area, playing fields and two golf courses, not unlike Sunken Meadow and other large public-use state parks already in the system. The plan also included a parkway from Southern State Parkway and Northern State Parkway, northward through historic Cold Spring Harbor and Lloyd Harbor, to connect to the park. The state acquired the rights-of-way to create the parkway in the 1960s, but it was never built due to local opposition and a statewide fiscal crisis that froze funding.

The property sat unused for fifteen years while New York State grappled with a recession and a massive debt crisis in the 1960s and 1970s. Some of

Farm building, Caumsett State Park. *Gail Frederick, licensed under CC by 2.0.*

the original structures on the Caumsett property remain today. The former polo pony barn is now an equestrian center, and one of the former garages houses an outdoor learning and environmental education group. Several other buildings on the property were demolished. Perhaps one of the most glaring omissions from the estate is the mansion's entire left wing. Field III's wife removed it in the 1950s to make the footprint of the main house smaller. However, it left the John Russell Pope–designed mansion glaringly asymmetrical. The two Lloyd houses at the entrance to the property also survived, and both currently house local historical and preservation groups.

Caumsett State Park opened in 1977 as a passive-use facility, mostly used for hiking, biking, birdwatching and horseback riding. The house was placed in the National Register of Historic Places in 1979. The Caumsett Foundation was established in 1995 to help restore and maintain the extant buildings and conduct educational, cultural and environmental programs in the park.

Today, the 1,400 acres of acquired property largely remain as a preserve. Extending to Long Island Sound, the park features miles of scenic paved pathways winding past the original farm buildings and stables to the grand mansion near the shoreline. Hiking trails extend down to the undeveloped beach and salt marsh and loop around through vast woodlands and meadows. Visitors can get a feel for what life was like on a former Gold Coast estate built by the heir of a vast retail fortune.

COLD SPRING HARBOR STATE PARK

Cold Spring Harbor State Park, located just west of the restaurants and small shops in Cold Spring Harbor's business district, has steep, rocky terrain that rises sharply from a small parking lot across from the scenic harbor on Main Street.

The placement of the park within the boundaries of a downtown area is unusual. However, what is even more unusual is the original use the New York State Department of Transportation planned when it acquired the property as a right-of-way in the 1960s.

The state acquired the piece of land, along with several other properties to the south, to develop a four-lane parkway to bring visitors to Caumsett State Park. The plan was to start the parkway as an exit from the Southern State Parkway in South Farmingdale. The state constructed a small part of this parkway, now known as Bethpage State Parkway, which currently terminates at Bethpage State Park. Caumsett State Parkway would have continued north from there, bisecting Northern State Parkway and Route 25 in the Woodbury area, and then proceeded north to Cold Spring Harbor. From there, the parkway would have traveled through Lloyd Harbor, bisecting West Neck Road and running east of the Immaculate Conception Seminary, where it would have crossed the harbor directly into Caumsett State Park. At that time, the state planned to convert the grand mansion and estate at Caumsett into an active-use park with a boardwalk, a beach, picnic grounds and a golf course.

It is easy to imagine how drastically the character of this quiet North Shore hamlet would have been altered if a parkway had been built through it. The parkway proposal was defeated thanks to the efforts of local residents and civic leaders who also managed to put a stop to the major redevelopment of Caumsett State Park. Today, that park and its mansion and grounds remain largely undeveloped, serving as a peaceful preserve.

The Cold Spring Harbor area has a rich and storied past. Originally occupied by Native Matinecock, the land was purchased by English settlers around 1653, and it became a center of commerce, industry and trade. It was initially called Cold Spring due to its abundant freshwater springs. However, its name was changed to Cold Spring Harbor in 1826 to differentiate it from Cold Spring in Upstate New York.

Early on, a gristmill was constructed, along with a sawmill for processing cut timber from heavily wooded areas. The area's other activities included farming, fishing, boatbuilding and nearby sand mining and brickmaking. Transport ships lined the harbor, bringing produce and other goods to the markets of New York City.

Steps from the state park is the historic hamlet of Cold Spring Harbor, New York. *Dougtone, licensed under CC BY-SA 2.0.*

In the mid-1800s, whaling became a significant industry in Cold Spring Harbor. The deep harbor port was an ideal location for bringing in whaling ships laden with oil, which, at the time, was used in lamps for lighting. The Cold Spring Harbor Whaling Company, founded by the Jones brothers, operated between the mid-1830s and 1860. The company once had nine whaling ships and employed half the town. When the sailors returned from months at sea, they were ready to drink and carouse at the bars and inns that lined Main Street. It was at that time that Main Street received its nickname: Bedlam Street.

Kerosene began to replace whale oil as cheaper alternative in the mid-1800s. In the late 1880s and 1890s, Cold Spring Harbor was transformed from a whaling port into a summer resort area, with grand hotels and resorts welcoming the upper echelon of society. Some of its more glamorous resorts included Glenada, which offered its guests yachting regattas, dancing, music, a casino and numerous dances and balls. As Cold Spring Harbor's popularity increased among the resort crowd, Mr. C.A. Gerard, a proprietor, opened nearby Laurelton and Mountain House to accommodate additional guests. His wife ran their Forest Lawn resort property. Mr. George F. Teal operated a hotel called Evergreen House at the mouth of the harbor. The area remained a popular resort until around World War I.

On an 1873 map of the area published by Beers, Comstock and Cline, the property where the current state park meets the main road and the harbor is shown as belonging to N. Seaman, with the G. Van Austell Hotel next door. The Seaman family operated the dock where the Seafarers Club was established in 1966, along with a general store. There is a small historic house that remains between the state park and the current library; it is called a Seaman house, but it is not the original Noah Seaman house. That house stood in what is now the state park's parking lot, according to Huntington town historian Robert Hughes.

A small housing development was once located to the west of the present-day library. Terrace Drive connected to a road called Harbor View, where there were several houses. When the state acquired the land to create Caumsett State Parkway, it demolished the Harbor View houses. It is possible that some remnants of the neighborhood's foundations still exist in the nearby woods.

Fiscal constraints and community opposition caused the state to abandon the proposed Caumsett Parkway project, and the land acquired for it remained unused for almost four decades. However, in the year 2000, the state became aware of the local library's need for a new location and

A photograph showing the houses that once stood on Terrace Place and Harbor View Road in Cold Spring Harbor, 1940. *Huntington Historical Society.*

agreed to allocate a portion of the land for the library's construction. The remaining land was dedicated as a state park, and it now serves as the northernmost connection to Trail View State Park, which is the former right-of-way for the parkway.

Today, you can drive down Main Street in Cold Spring Harbor and park in the small lot. Cold Spring Harbor State Park offers a challenging trail with rustic steps that lead up steep bluffs through the woods. A hike to the top is rewarded with a stunning view of Cold Spring Harbor, which is filled with numerous fishing and private recreational boats. After your vigorous hike, you might need refreshment. You can head into the historic town, where there are many wonderful restaurants and one-of-a-kind shops.

11

CONNETQUOT RIVER STATE PARK PRESERVE

The dark and imposing wood-shingled mansion that stands at Connetquot River State Park in Oakdale was once a grand clubhouse for wealthy members of the South Side Sportsmen's Club, who built on the site of a small 1800s stagecoach inn once called Snedecor's.

The Native Secatogue named the place Connetquot, meaning "at the great long river." The river here meanders for over ten miles, fed by underground springs that originate in the Islandia/Ronkonkoma area. It runs south through the park and continues until it empties into the Great South Bay. In 1683, William Nicholl acquired fifty thousand acres of land from Sachem Winnequaheagh and became the area's first patentee. In the early 1800s, Eliphalat Snedecor purchased three hundred acres of land for farming from the Nicholl family. A stagecoach stop was located nearby, and Snedecor saw an opportunity to create a small inn for travelers. It was known as Snedecor's Inn, or Snedeker's Hotel, and later, it became the Connetquot Hotel when Snedecor expanded it in the 1830s.

The inn became increasingly popular with wealthy Manhattan sportsmen who enjoyed the rustic surroundings and outdoor sporting activities, along with Snedecor's renowned mint juleps and champagne punch. In 1865, a group of prominent New Yorkers approached Snedecor's son Obadiah, who had inherited the inn, to see if he would sell it to them for private use. Snedecor agreed to sell, and in 1866, the South Side Sportsmen's Club was established on what would later become Connetquot River State Park. Membership in the club was limited to one hundred men, who pooled their

The original Snedecor's Inn, with a covered porch, can still be seen on the far-right side of the former South Side Sportsmen's Club building. *Long Island State Parks.*

resources to purchase and maintain the club. At first, the club remained rustic. The club added a few amenities to Snedecor's original inn, including a billiard room, dining facilities and bedrooms for members. Massive fireplaces and stoves heated the building, and oil lamps lit the rooms. The members' activities included duck and pheasant shoots, fishing and relaxing. Many members enjoyed the peace and quiet and the opportunity to fraternize with like-minded companions.

In 1886, the club expanded to accommodate more bedrooms, dining rooms, a porch, a large kitchen and other comforts. These amenities included a three-story addition with a forward-facing two-gabled roof. The club left the original Snedecor's Inn intact and simply built the extensions around it. If you stand facing the front of the sprawling clubhouse, you can see the original Snedecor's Inn on the far right side of the building. It is lower in height than the later additions and clearly holds its early American character. The club likely saw no reason to knock down the original, solidly constructed inn, and thus preserved a time capsule of Long Island's early history within the larger clubhouse structure.

While the additions vary in height, the clubhouse has the same type of dark wood shingles throughout the exterior of the building, so there is a level of unity in its façade. Inside, heavy wood paneling and molding create a dark, sophisticated, rustic feel, which would have seemed cozy when the fireplace was blazing. There were plenty of rocking chairs situated around the fireplace area, and there were outdoor gathering areas where members could relax in the cool, fresh air and exchange stories about their hunting and fishing adventures. Members also had their own lockers to store their booze and other important items they wanted to leave for their next visit.

The roster of former members reads like a who's who of New York's high society. In the club's heyday, its members included such luminaries as William K. Vanderbilt, August Belmont, Louis Comfort Tiffany, artist William Sidney Mount and William Bayard Cutting, along with other well-known industrialists, politicians and artists. The club permitted members to bring invited guests, and some of these included presidents Theodore Roosevelt and Ulysses S. Grant, and Andrew Carnegie (once the richest man in the world). Even later, toward the end of the private club's era, its members included executives from some of the largest corporations in America, including PepsiCo, American Airlines, Westinghouse, Getty Oil and others.

Over the years, the club acquired another three thousand acres from nearby property owners. In 1900, a separate annex building was constructed to provide extra accommodations for members and their guests in more luxurious and private quarters. Its exterior, dark-wood, shingled style is similar to that of the grand clubhouse, as is the architecture of its horse stable, which was built before there were automobiles. Back then, members would either take their own horse and carriage to the club or ride the railroad to the nearby station, where the club would send a carriage to welcome them.

Other notable structures that are still extant on the site include the Nicholl Grist Mill, which dates to the earlier original patentee's settlement. The mill contains a unique Scandinavian mechanism that is quite rare to find in the United States. There is also a circa-1800s icehouse on the property that was built prior to refrigeration. During the winter months, workers cut ice out of local freshwater streams and stored it underground in the icehouse, where it would typically last well into the summer months.

The club also created a fully functioning fish hatchery that still exists. Revolutionary for its time, the hatchery was used for breeding and growing fish that could be used to stock the rivers and streams on the property. The hatchery is located about a mile from the main clubhouse, down a well-

The Nicholl Grist Mill, still extant on the property. *Author's collection.*

marked trail through scenic woods and fields. It was built in 1891, although there were earlier efforts to propagate fish downstream from the site. Today, the state still uses the hatchery to stock the Connetquot River, along with the nearby Nissequogue River.

Into the early to mid-1900s, club members still enjoyed the pristine wilderness setting, even as developers in the surrounding area began building housing and other structures of industry. Increasingly, the cost of maintaining the club and its property taxes became burdensome, and in 1963, the club agreed to sell its land to New York State for $6.2 million. Part of the agreement stipulated that club members could lease back 583 acres for their private use for ten years. In 1973, at the end of the agreement, the state converted the property into a public park.

Connetquot became New York State's first state park preserve. The "preserve" designation recognizes the value of maintaining the ecological, cultural and historic resources found on the property. Visiting the site is like stepping back in time, to when the South Side Sportsmen's Club used the property.

The state maintains 3,473 acres of land and water for the protection and propagation of game birds, fish and animals. Aside from its historic structures, the park is home to deer, raccoon, game birds and migratory birds, along with numerous native trees and plants. Anglers come from all over to experience the park's freshwater fly-fishing (by permit only).

There are fifty miles of trails for hiking, biking and cross-country skiing winding through the park's scenic woods and fields. Interesting and informative kiosks created by the Friends of Connetquot are scattered throughout the property and explain the historical and natural history of the site. Visitors can also attend regular tours and programs inside the historic clubhouse, where a small group of sportsmen once gathered to enjoy the rustic charm of the site.

12

GILGO STATE PARK

While no one is certain where the name Gilgo comes from, one story says that the Burch family, who lived on Long Island's South Shore, had a son named Gil who would go fishing along this stretch of barrier beach. If neighbors were going to fish there, too, they would say they were going where "Gil goes."

It's likely that Gil knew—as others do today—that Gilgo is a great fishing spot.

While Gilgo State Park is located off Ocean Parkway in Babylon, many people may not even be aware it exists, since it is accessible by permit only. It is also confusing, because the state park shares a name with a Babylon town park and a small community nearby, both called Gilgo.

In the 1800s, this stretch of barrier island around Gilgo State Park was mostly deserted, except for a string of United States Life-Saving Service (USLSS) stations the federal government erected every three miles or so between Rockaway Point and Montauk. The USLSS stations patrolled the beaches and kept an eye out for ships in distress. Members of the USLSS, the predecessor of the U.S. Coast Guard, were known for their bravery in the face of harsh storms and extreme conditions during their many shipwreck rescues.

Around the mid-1880s, there were accounts of ferryboats bringing visitors to shore near Gilgo State Park. A small community started to form around an area called Hemlock Beach, west of the current state park. There was a USLSS station and a small community there. A Mr. Wesley Van Nostrand had a pavilion where he would serve meals to visitors who could then relax or walk or the ocean beaches before returning home. At the turn of the twentieth century, several *Southside Signal* newspaper articles offered accounts of guests enjoying music and dancing at an expanded hotel at Van Notrand's.

Nearby, Wessel's Hotel also accommodated visitors to the ocean shore. A 1914 advertisement for Wessel's Hotel in *South Side Signal* promises visitors an easy ferry ride from Babylon and Lindenhurst to experience fantastic boating, bathing and fishing. The hotel offered overnight stays by the day or week in addition to dining and weekend parties.

Unfortunately, only three months after the Wessel's advertisement, a December nor'easter brought pounding surf and flooding to the area. While there are records of Van Nostrand's still in existence afterward, the beach area was greatly compromised, and several buildings in the community around Hemlock Beach were destroyed. In the next decade, many of the surviving cottages were moved away from the shore, north of what is now Ocean Parkway, to communities in the Town of Babylon's Gilgo Beach and West Gilgo Beach.

Those communities were likely quite concerned when, in 1928, the Long Island State Park Commission announced its plans to create an Ocean Parkway through the center of the island. These isolated hamlets, previously accessible only by ferry, now faced the prospect of a four-lane parkway running the entire length of the barrier island. Between 1928 and 1930, the park commission acquired land from Oyster Bay, Babylon and Islip Townships, along with numerous other private property owners, to construct the parkway from Jones Beach to Captree State Park. Many residents opposed the plans, likely fearing large-scale development and throngs of visitors. However, when the state obtained the property rights to build Ocean Parkway, it also acquired much of the land north and south of it, from the bay to the ocean. This actually worked to keep it out of the hands of large-scale developers, so aside from the Ocean Parkway and some preexisting small communities, the barrier island's bay and oceanfront are still pristine, with no high-rise hotels, throngs of tourists or amusements.

The Long Island State Park Commission acquired 464 acres from the Town of Babylon to create Gilgo State Park in 1928. However, it did not create a massive public park beach like the ones found at Jones Beach. Instead, in a seemingly unusual move, the commission proposed consolidating the U.S. Coast Guard operations into a centralized station at Gilgo State Park, in effect, changing how the coast guard and its predecessor, the USLSS, operated in the area. Instead of having a station every three miles, the United

The coast guard station at Gilgo State Park. *Long Island State Parks.*

States government agreed to take down several of the former stations and appropriate $100,000 for the construction of a new coast guard station at Gilgo State Park. This way, the state was able to build its Ocean Parkway unimpeded by previous lifesaving operations in the area.

The new coast guard station at Gilgo State Park was located near the current park entrance. It was an imposing, two-story brick structure with a glass-enclosed lookout tower. A 1932 article in the *New York Times* described the new station as being "modernistic in design" and conforming in architectural style to the buildings at Jones Beach State Park. The central location saved the federal government money, since the new Gilgo station would patrol fifteen miles of shoreline. A boat basin, which still exists on the north side of the park, was built to connect with the New York State boat channel and provide access to the Great South Bay. During World War II, this station was part of a contiguous chain of East Coast stations that kept watch and patrolled the coastline for any potential saboteurs or attackers. They scanned the shore on foot or on horseback and protected the United States from potential invasion.

After the 1940s, the station at Gilgo State Park was decommissioned, and operations were consolidated at Coast Guard Station Fire Island no. 83, which still exists on the north side of Robert Moses State Park.

Gilgo State Park is located just west of the Town of Babylon's Cedar Beach. It is an undeveloped preserve accessible by only four-by-four-wheel-drive vehicles with permits. The only remnants of the coast guard station that existed there were removed in the 2020s. The park comprises 1,200 acres that front the Great South Bay to the north and the Atlantic Ocean to the south. Its tranquil shoreline features fantastic white sand beaches and spectacular views. Surfcasters know it well as a spot for striped bass and bluefish in season, as Gil knew. Swimming is not permitted, as there are no lifeguards. Stewards regularly patrol the beach to manage the population of federally protected piping plovers that nest there. Visitors are asked to respect the flora and fauna of the park, including important native plant species that grow in this particular area.

HALLOCK STATE PARK PRESERVE

Hallock State Park in Riverhead is named for the Hallock family, who were part of the original Puritan settlers who came to the area in the mid-1600s. William Hallock settled on the land now called Hallock Lane in 1661, and while other families, including the Petty, Horton, Swezey, Wines and Cooper families, were part of the original land grant, Captain Zachariah Hallock later acquired most of the land in the area before he died in 1820.

The area housed so many Hallocks—fourteen homesteads at one point—that people in the area just took to calling it Hallockville. Through nine generations of Hallocks, many of the sons in the family were given a piece of the land here to farm. While many of the original Hallock homesteads no longer exist, around half a dozen remain, and there are still Hallock descendants who live nearby. At the Hallockville Museum Farm, located just west of the Hallock State Park entrance, visitors can tour some of the preserved Hallock houses and farm buildings to learn more about their history and get a feel for what life was like in Hallockville during that period.

Captain Zachariah Hallock was a member of the minutemen, who were a part of America's earliest militias. They were called the minutemen because they had to be ready "in a minute" to fight against any threat. During the Revolutionary War, it is believed Captain Hallock remained in the area, despite the British occupation of Long Island, according to Hallockville history. During the War of 1812, Captain Hallock's son

The Hallock homestead at the entrance of Hallock State Park. *Author's collection.*

Zachariah Jr. played a role in a skirmish with the British just offshore from the current state park, according to a detailed account researched by historian Richard A. Wines. In his book *Defense of the Eagle*, Wines describes how Zachariah Hallock Jr. discovered a cutter ship named the *Eagle*. It had come under heavy attack by a British ship that had had rammed it to the shore. Hallock Jr. sent his ten-year-old son, Herman, running to alert local militia members, who came to the aid of the *Eagle*. They fought off the British for three days until, ultimately, they were outnumbered and outgunned, although it was a brave attempt.

As the Hallocks and other early settler families continued to work the land over the centuries, in the early 1900s, a large influx of Eastern European immigrants, including the Cichanowicz, Kujawski, Trubisz, Naugl, Romanowski and Celic families, came to work the land, according to Hallockville Museum Farm. They started out as immigrant laborers on the land; however, they went on to work diligently and save up enough money to buy their own farms in the area.

Around this time, the Boys' Club of New York began looking for a place in the country where they could bring underserved boys from New York

City's poor neighborhoods to experience nature and the outdoors in a camp environment. In 1902, the club acquired thirty acres in what is now Hallock State Park Preserve and later expanded it to hold one hundred acres in the 1940s. The club called the area Camp William Carey, named after the Boys' Club trustee who developed the first summer camp program for the organization.

The camp provided young people from Manhattan a much-needed respite from the summer heat and exposed them to fresh air, exercise and recreation. In a 1927 letter to the editor, Boys' Club president Richard Sabin wrote to the *New York Times* that the camp on Long Island appears to "do as much for their minds as it does for their bodies" and described the positive influence Camp William Carey had on the children who visited.

The camp operated on this site for the next sixty years. While the camp is now closed, the Boys' Club still exists and promotes social, physical, intellectual, vocational and professional growth opportunities for New York City youth. In the 1950s, this author's father, Jerry Matejka, served as a counselor at Camp Carey and wrote an account of his time there, which he donated to Hallock State Park Preserve. In his recollection, he states:

In 1957, while attending Stuyvesant High School in Manhattan, the Boys' Club hired a small group of fellow classmates and me for an eight week employment as camp counselors. I also remember other fellow counselors were hired from prep schools as far away as Massachusetts. We were paid less than $800 for a full summer's work, but it was a rewarding experience. Every two weeks, a new group of boys aged ten to fifteen, mostly from the Lower East Side of New York City, would arrive for their vacation. The camp had a large recreation hall, which was where counselors served all meals to their assigned boys. The accommodations consisted of cabins that housed about twelve to fifteen boys and the counselors. During the day, besides baseball, volleyball and basketball, the campers had two choices for swimming. There was a steep staircase leading down to the beach on the Long Island Sound. Otherwise, there was the pond a short walk away, which was also very enjoyable for the campers.

This was a self-contained facility, so the campers were on the premises for their entire stay. For the counselors, it was pretty much the same. However, several times, we hiked from the camp to Mattituck after lights out at 8:00 p.m. to try to catch the last show at the local movie theater in

Camp William Carey counselors, including the author's father, Jerry Matejka (*top row, fourth from right*), 1957. *Photo provided by Jerry Matejka.*

> *town. That theater is also long gone and is now a sporting goods store on the Main Road (Route 25).*
> *These were wonderful and cherished memories of an earlier Long Island.*

When the camp was in operation, it cleared the land around Hallock's Pond and placed a lifeguard on duty so the boys from the camp could swim. Many of them preferred the smooth sand around the pond to the rocky beach of the Long Island Sound, which, many North Shore swimmers will attest, likely hurt their feet. Both the beach and the pond are still accessible from trails in the current state park, although the area surrounding the pond is heavily wooded and swimming is not permitted. Camp Carey ceased operations in 1963, and the camp buildings were removed in the decades after. One of the only structures remaining is a small concrete and brick incinerator used by the camp that still stands in the woods.

Following the closure of the camp, Levon Corporation/Curtiss Wright acquired the property along with another four hundred adjacent acres of farmland, where it planned to build an industrial complex. The complex never appeared. Instead, the land was heavily cleared and mined for sand, which was used in the construction of buildings and roads in New York City and Long Island. Sand mining was as lucrative as it was irreversibly damaging to the environment.

In the early 1970s, Levon Corporation/Curtiss Wright closed down operations after several lawsuits and protests by local residents. You can still see the effects of the sand mining operation along the path that runs down to the beach. In one section, an entire bluff that was once over one hundred feet tall and all its vegetation are missing. All that remains is a large sandy crater where very little will ever grow.

In 1973, Long Island Lighting Company (LILCO) acquired the five-hundred-acre parcel to build four nuclear power plants. Despite years of protest, the U.S. Nuclear Regulatory Commission (NRC) initially gave approval for a 1,150-megawatt nuclear power plant. However, the opposition gained ground, and in 1980, the project was officially defeated. LILCO was the predecessor of Key Span Energy, known today as National Grid. The utility company is charged with providing electricity to Long Island residents.

In 2003, the New York State Office of Parks, Recreation and Historic Preservation (formerly the Long Island State Park Commission) acquired the land from the Trust for Public Land. With assistance from Key Span Energy, Peconic Land Trust, the Town of Riverhead and the Long Island Farm Bureau, the commission was able to create a unique arrangement that combines a public park with preserved farmland.

The state park preserve comprises 225 acres with trails that wind through the woods, where some of the North Fork's oldest families once lived and farmed and children from the inner city may have experienced the beauty of nature for the first time. The nearly one-mile-long sound-front beach is accessible via marked trails. Very large glacial erratic boulders that were deposited here at the end of the last ice age are scattered along the shoreline. Native flora and fauna fill the park. You can see native white yarrow, purple bergamot and native grasses, along with wild raspberry, white oaks, silver maples, sumac, eastern red cedars and more. It is home to woodpeckers, catbirds, sapsuckers, turkey vultures and blue-winged warblers, along with numerous species of turtles that live in Hallock's Pond.

The state originally dedicated the park as Jamesport State Park in 2005. Five years later, the parks department created a master plan for the park, at which time it changed its name to Hallock State Park Preserve. The park features a visitor center, gift shop, small exhibit area and pleasant outdoor seating area.

The Trust for Public Land sold the development rights for the three hundred acres of land in front of the park on Sound Avenue. By doing so, the trust preserved those acres of farmland in perpetuity through a conservation easement. It divided the land into eight separate parcels that it sold at affordable prices to eight farming families. The farm owners are free to work their land but can never develop it for housing or other intensive use, thus maintaining the historically rural character of the area.

14

HECKSCHER STATE PARK

The vast, rolling fields and acres of woodland at Heckscker State Park in East Islip were once home to two extraordinary mansions that were both occupied by a series of characters whose unusual stories made local headlines at the turn of the twentieth century.

However, the park's history dates to the earliest founding of the area, when William Nicholl, considered the founder of Islip, came to the New World with his father, Matthias, in 1664. The Nicholls were part of a British contingency sent here by King Charles II's brother James, Duke of York, to assert English control of Long Island over early Dutch settlers. When the Dutch quickly receded, Matthias Nicholl was appointed first secretary of New York, named for England's Duke of York, as his just reward. Matthias was involved with the creation of the Duke's Laws, which established the system of courts and trials for the new settlements. In the latter half of the 1600s, Matthias also went on to become a judge, a speaker of the assembly and the mayor of New York. William Nicholl, following in his father's political steps, became a New York attorney general and New York State assemblyman.

In 1683, William Nicholl decided to settle in this rich and fertile portion of Long Island situated along the Great South Bay. He acquired over fifty thousand acres on the eastern side of what is now the town of Islip and built his house on the land where Heckscher State Park is now located. He purchased the land from the Secatogues and Sachem (or Chief) Winnequaheagh of Connetquot.

His British connections easily secured him a royal patent for the land, solidifying his claim. Nicholl created a plantation called Islip Grange there. He named the town after Islip in Northamptonshire, England, where he and his father had previously lived. William Nicholl died in 1723.

In the decades that followed, the land around Islip Grange was used for agriculture, cattle grazing, fishing, boating and shipping. However, by the late 1800s, wealthy industrialists from New York City had begun traveling to the area to take advantage of its abundant sporting opportunities. They formed private sportsmen's clubs in the area, most notably the South Side Sportsmen's Club at what is now Connetquot River State Park. These elite visitors sought a retreat from the sweltering summer heat of the city where they could hunt, fish and relax. Some also established their own summer mansions.

In the late 1800s, Edwin A. Johnson acquired several hundred acres of land in the area called Deer Run Farm, now part of Heckscher State Park. Johnson was a wealthy New Yorker who chose to retire in Islip, presumably to be near the South Side Sportsmen's Club. He was one of the oldest members of the club when he passed away in 1933, and in fact, he chose to live at the club for the last several years of his life.

Sarah Ives Plumb and James Neale Plumb acquired Deer Run Farm well before Johnson's death and established it as a game preserve. The deed to the property may have been in both their names; however, it was Sarah Ives's money that purchased the estate. She was made quite wealthy from a family inheritance. Unfortunately, Sarah Ives Plumb died in 1877, just five years after the purchase of the nearly seven-hundred-acre Deer Run Farm in Islip. Her grief-stricken husband donated funding to Emmanuel Episcopal Church in her memory and dedicated a stained-glass window in the back of the church as a tribute to her. The church still stands in Great River.

However, not soon after her death, Plumb began making fantastical additions to the house, including a four-story central tower, numerous gables, arches and highly ornate and whimsical trim work. It was quite the sight to see. His daughters, Mrs. Marie Jeannette Plumb Nares and Miss Sarah Lenita Plumb, accused him of squandering their inheritance. A judge cleared Plumb of the charges in 1895. However, Plumb developed a deep and unrelenting hatred of Alexander Masterton, the attorney Plumb's children hired when they accused him of diverting money out of their trust fund. In 1899, Plumb shot Masterson dead and freely admitted what he had done, afterward acting quite satisfied with himself, according to reports at the time. He died later that year in Bellevue (psychiatric) Hospital.

Sarah Ives Plumb and James Neale Plumb's estate. *East Islip Historical Society.*

The former Plumb mansion, greatly reduced, was moved to the south side of West Main Street in Bayshore, where it still stands. *Author's collection.*

Plumb's son, James Ives Plumb, inherited the estate and promptly got out from under it by selling it to George C. Taylor in 1903. Taylor already owned eight hundred acres to the west of the Plumbs and envisioned creating a grand estate that combined the two properties. He proceeded to deconstruct the once-fabulous Plumb mansion, taking down its grand towers and gables and moving it to the south side of Main Street in Bayshore. Interestingly, the much-subdued former Plumb mansion still stands today with boutique shops on its lower level. It is nearly unrecognizable from its original grandeur. A careful eye will notice the trim work and two front gables that hint at the home's former identity. Back at Heckscher State Park, the only visible remnant of the Plumb estate is a row of trees that once led up to the Plumb mansion.

According to many reports from the time, George C. Taylor was a wealthy eccentric worth more than $20 million. After merging the Plumb property with his own, Taylor set up a farm and preserve with thirty outbuildings, including stables, dairy barns, greenhouses and servants' quarters. The preserve included a herd of wandering elk. The Taylor mansion was not as ostentatious as the Plumb mansion, yet it had its own grandeur. It was three stories high with a wraparound porch on the first two levels. It contained no gables or dormers, being purely linear in design except for the two-story portico and colonnades at the front door.

Taylor never married; however, his housekeeper and companion, Betsey Head, and her daughter, Lena, lived with him at the estate. He was apparently quite fond of Lena and provided her with all she could need, including private tutors and instructors. It was assumed that Betsey—and eventually Lena—would become the heirs of Taylor's grand estate.

However, everything changed when Lena Head fell in love with one of Taylor's workers, a landscape gardener named William F. Bodley. Against her mother and Taylor's wishes, Lena eloped with him in 1901; she was eighteen, and Bodley was thirty-five. Taylor was outraged, and he and Betsey threatened they would not leave her one single dollar of the vast estate. But Lena had made up her mind. The newspapers at the time broadcasted the sensational story. Afterward, Betsey and Taylor stopped socializing altogether, and there were reports of heavy drinking and arguments at the estate. Betsey Head died in June 1907, and almost true to her word, she, in fact, left her daughter only five dollars. George C. Taylor died a few months later and left his estate to his nieces, nephews and other family members.

The mansion sat largely unused, and the family members who inherited it formed a corporation called the Deer Run Corporation to manage it.

George C. Taylor mansion. *East Islip Historical Society.*

They tried to sell it privately but with no success. In 1924, the officers of the corporation gave Long Island State Park Commission an option to buy the property for $250,000, pending a bond issue act. However, some neighboring estates in the area opposed the sale and the proposition of forming a state park on the land, as they feared an influx of crowds to the area. W. Kingsland Macy of the nearby Timber Point Country Club and Horace Havemeyer, his brother-in-law, offered to buy the estate at the same price the state was offering. Deer Run agreed to the sale. The property was sold to the newly formed Pauchogue Land Corporation, which then joined with wealthy landowners to the north to block not only the state park but also the proposed Northern State Parkway route through their estates, according to Chester Blakelock, who kept a record of the proceedings in his 1959 *History of the Long Island State Parks.*

However, the Long Island Park Commission was determined to acquire the Taylor estate and sought eminent domain. A multiyear lawsuit ensued. After considerable delay, the state prevailed in 1929. However, it no longer had the dedicated funds to make the purchase. August Heckscher, a wealthy philanthropist who believed in the public benefit of cultural and recreational facilities, stepped in and offered $262,000 toward the state's purchase of the property. Heckscher was a wealthy industrialist and real estate investor whose interests turned to philanthropy later in his life.

He was most interested in uplifting and bettering the conditions of New York City's poor by dedicating parks and playgrounds for their use and enjoyment. The state was originally going to name the park in East Islip Deer Run State Park; however, in recognition of Heckscher's philanthropy, the state named it Heckscher State Park.

The state had originally planned to use the Taylor mansion as state park headquarters, but before the lawsuits ended, the Long Island Park Commission had already established its regional offices at Belmont State Park and no longer needed the mansion. The state demolished it in 1933.

At the park today, there are spectacular views of the Great South Bay and Fire Island. The park offers opportunities for bay swimming, boating, sailboarding, kayaking and canoeing in designated areas. Newly built waterfront cabins that can accommodate two to six people are available for three-season rental and include a living room, bedrooms, full kitchen, bathroom and screened porch.

Not much remains of the former estates, except the wide-open, undeveloped land. An original barn is still standing that park maintenance uses as a garage, and a former carriage barn from the estate was converted for recreational use. Visitors to the 1,600-acre Heckscher State Park can enjoy its large shady groves of trees, ideal for family and group picnic outings. Grills are available. There are also open fields for soccer, cricket, lacrosse and other sports. The park has 4 miles of paved hiking and biking trails and is the southern starting point of the 31.8-mile Long Island Greenbelt Trail. Dogs are permitted but only in undeveloped parts of the park at the west end of parking field five.

15

HEMPSTEAD LAKE STATE PARK

The most prominent feature of Hempstead Lake State Park in West Hempstead is, of course, the massive lake that shares the park's name. Hempstead Lake covers 167 acres and is nearly three miles in circumference, making it the largest lake in Nassau County.

Hempstead Lake would have looked quite different centuries ago. The Hempstead Plains once covered this area from the Queens border to Suffolk County. Back then, high grassland stretched as far as the eye could see. Natives likely knew about the fresh water here. However, the lake, at that time, probably would have been marshier and more swamp-like. Nearby, the Natives named an area Merrick, meaning "plains country" or "grassland." They also used the term *Massapequa* to describe a "land with fresh water." Both names still exist in the names of nearby towns.

The name Hempstead comes from Hemel-Hempstead, which was the name of the town the first English settlers to the area came from in Hertfordshire County, England. In addition to being tied to the English, Hempstead's early history is intertwined with that of early Dutch settlers to the west and the Natives of the Rockaways who occupied the land between Queens and southern Hempstead.

The first English settlers included Reverend Robert Fordham, who negotiated with sachems in the Rockaway, Massapequa and Merrick areas to acquire most of the Hempstead Plains. The Dutch approved the patent for the land in 1644. A settlement, semiautonomous from the Dutch, grew around current Hempstead Village. Twenty years later, the British came to Long Island under the direction of the Duke of York to wrest control from the Dutch, asserting their dominance over the land they then called New York.

Early settlers used the land in Hempstead mostly for farming and grazing over the centuries that followed. At the onset of the American Revolution, Loyalists in the southern area of Hempstead aligned themselves with the British, while the northern portion of the town supported the ideals of the Patriots. It is the reason why, to this day, the town is divided into North Hempstead and (south) Hempstead. The sides would have been fierce opponents during the Revolution. Interestingly, Hempstead Lake is mentioned in a Revolutionary War historical novel *Oliver Wiswell*, by Kenneth Roberts. In that book, Roberts describes hundreds of British Loyalists using the densely covered, swampy area around the lake as a hideout.

After the war, the area around Hempstead grew modestly into the 1800s. However, the nearby city of Brooklyn continued to grow exponentially. Not yet a borough of New York City, the City of Brooklyn acquired the land that included Hempstead Lake to serve as a water supply as it faced a growing crisis in the mid-1800s. At that time, it dammed the Mill River on the site to form the current reservoir of fresh water. The dam's outlet

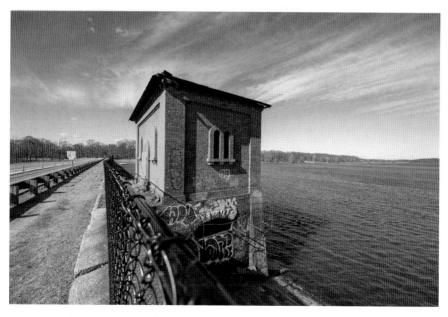

Outlet controls for the dam, built for the City of Brooklyn, were housed in a gatehouse-like structure, still extant on the property. *Long Island State Parks.*

controls were housed in a gatehouse-like structure, still located on the property, that was constructed to direct water through a brick pipe arch from the dam into South Pond.

Community pressure against Brooklyn's plan to tap into Long Island's reservoirs arose in the late 1870s. The metropolis was also eying other Long Island water reservoirs as far east as the Pine Barrens in Brookhaven to service its growing population. However, the plan was becoming a behemoth. The costs were far exceeding the budgeted $1.4 million allotted to bring the reservoir up to a planned billion-gallon capacity. In 1898, the crisis was averted when Brooklyn consolidated with the greater City of New York. It was then able to tap into New York City's water supply from Upstate New York.

The dam and the lake it created sat dormant until the 1920s, when the Long Island State Park Commission requested that New York City dedicate the reservoir and its surrounding acres for public use. The *1930 Land Acquisition Annual Report* for the Long Island State Park Commission shows the City of New York dedicating 2,200 acres in 1925 with another 3.5 acres of woodlands purchased from Fredericka Kosel in 1930. The dedication, in the form of a surface easement, included present-day Hempstead Lake State Park, Valley Stream State Park, Massapequa Preserve and important acreage for portions of the Southern State Parkway, Meadowbrook Parkway, Wantagh Parkway and Bethpage Parkway.

At its opening, Hempstead Lake State Park had tennis courts, boating, trails for horseback riding and a miniature golf course. Visitors came from the city to spend the day picnicking or hiking in the woods. Native oak, birch, beech and pine trees surrounded the lake, which was stocked with bass and trout. One of the park's most beloved features was a hand-carved carousel built around the turn of the twentieth century by M.C. Illions and Sons Carousel Works of Coney Island. August Heckscher, the well-known Long Island philanthropist, donated the carousel to the park in 1931.

In the mid-1900s, the population around the park grew rapidly, and dense suburban housing, highways and shopping centers sprang up around it. Pollution became an increasing concern. In the 2020s, New York State embarked on a historic $47 million improvement plan to help mitigate runoff and pollution risks at Hempstead Lake State Park. The upgrades, completed in June 2023, included one of New York State Parks' largest-ever wetlands restoration projections. The state allocated nearly $17 million in funding to mitigate flood risk on the Mill River and reduce the pollution entering Hewlett Bay at the 144-acre Northern Ponds complex. Eight acres

The control directed water through a brick pipe arch from the dam into South Pond. *Long Island State Parks.*

of new emergent wetlands allow water runoff from Southern State Parkway to be slowly filtered and purified before it enters Northeast Pond. The state also conducted restoration work to maintain the existing water level at the Northwest Pond Dam, creating a cleaner, more diverse area habitat. The project also included the removal of invasive species and the replanting of native species to support enhanced plant and animal biodiversity. Trash has been a consistent problem at the park. New systems were put in place to help prevent floating debris from entering the ponds area. Astonishingly, the state removed over one hundred tons of decades-old trash and debris that had come down the Mill Creek from the Northern Ponds complex, according to the New York State Governor's Press Office.

The massive environmental project also included structural improvements to stabilize the historic Hempstead Lake Dam, built in 1873. It is the only "high-hazard dam" on Long Island, meaning that a structural default could result in loss of life or property. The project further restored the dam's operational infrastructure to better maintain water levels and support the habitat and ecosystems of the park's waterways. The underground pipe arch culvert at Hempstead Lake gatehouse and South Pond inlet house was restored and is now equipped with new sluice gates that allow lake water levels to be controlled prior to and during storms. Water level monitoring and temperature gauge equipment installed at the gatehouse provide information about seasonal and long-term changes in the lake.

Many of the state's improvements focused on creating a healthier ecosystem at the park. However, it also made several upgrades to accommodate a greater number of visitors. A new observation deck was added at Northeast Pond, along with a two-mile stretch of Americans with Disabilities Act–compliant trails. The state expanded a greenway trail that provides a continuous north-to-south trail system through the park, along with a new eight-foot-wide wetland trail and two new pedestrian bridges. Other recently completed work at the park includes an eight-thousand-square-foot center that provides hands-on learning opportunities about storm resiliency and environmental management. New York State Office of Parks, Recreation and Historic Preservation; the Office of Resilient Homes and Communities; New York State Department of Environmental Conservation; United States Department of Housing and Urban Development; United States Army Corps of Engineers; United States Environmental Protection Agency; and United States Fish and Wildlife Service all cooperated on the project. It was, in fact, a massive undertaking.

Today, visitors to Hempstead Lake State Park can still see the dam that was once built to supply water to Brooklyn. There are bridle trails for horseback riding, biking and hiking and beautiful, shaded picnic areas that are a pleasure for friends and families gathering during the hot summer months. Hempstead Lake, McDonald Pond and South Pond are accessible for freshwater fishing and are stocked with black crappie, yellow perch, bluegill sunfish, pumpkinseed sunfish and chain pickerel, according to the State Department of Environmental Conservation. The park has eighteen tennis courts, six pickleball courts, playgrounds, basketball courts and a softball field, along with the historic carousel that has delighted families for nearly a century.

HITHER HILLS STATE PARK

The 1,755 acres of serene oceanfront beach and parkland surrounding present-day Hither Hills State Park in Montauk might have looked very different if a developer in the 1920s had succeeded in his plans for the area.

At that time, an entrepreneur and real estate investor named Carl G. Fisher envisioned creating a sprawling, luxury resort town in and around the Montauk area. He imagined building elegant hotels and a high-end casino that would attract wealthy patrons who would arrive here from as far away as Europe aboard luxury steamships that would dock in the harbor. Fisher, who created the densely populated Miami Beach years earlier, had a similar vision for Montauk. Today, only a few buildings and developments remain east of Hither Hills State Park, remnants of Fisher's ambitious plan.

Montauk is named for the Native Montaukett who lived here long before Carl G. Fisher and his big ideas arrived. They used the grounds for hunting and fishing and coexisted with the early English settlers who came here throughout the 1600s and 1700s to use its pastures for grazing cattle. The first house in Montauk was built in the 1740s, just on the opposite side of what is now the parking lot at Hither Hills. It was called Hither House, and a cattle keeper was stationed there to keep track off the wandering herds. Two other houses were built in the area and are now called Second House and Third House. Those original houses, also used for cattle keepers, were burned and then rebuilt and still stand east of Hither Hills at Kirk Park and Montauk County Park.

The Tower at Montauk once served as Carl Fisher's offices. *Katie Matejka.*

In 1879, the Town of East Hampton placed a large section of the Montauk area up for auction. The winning bid of $151,000 went to Arthur W. Benson of Brooklyn, who bought nine thousand acres to create his own private fishing and hunting retreat. Benson was a real estate developer and the president of the Brooklyn Gaslight Company. Years earlier, he had subdivided portions of Brooklyn to create Bensonhurst by the Sea, now simply called Bensonhurst, in the southwestern portion of that borough.

There were many disputes over the years related to Benson's deeded rights from the Montaukett and their relocation from the area. While Benson may have had broader ambitions for the use of the property, he built only a cottage on the property and left it largely undeveloped before he died in 1890. Austin Corbin, a wealthy businessman, banker and developer who also owned the Long Island Rail Road, purchased a portion of the land from the Benson heirs. Corbin's particular vision included building a deep-water port where ships from overseas could offload their cargo and have it transported by rail to New York City. By 1895, the rail to Montauk was complete, but just one year later, Corbin was thrown from a horse-drawn carriage and killed. His vision for a great shipping port at Montauk was never realized.

However, the rail brought more visitors from New York City to Montauk. The Montauk Inn, located in the same area as the current Montauk Manor, opened in 1899. By the mid-1920s, Carl G. Fisher had purchased a large swath of Montauk to create a mega-resort area among the high bluffs and shifting dunes of this scenic stretch of land on the eastern end of Long Island. In addition to hotels and a casino, the area would include a golf course, marina and business district.

Fisher grew up poor in Indiana, dropping out of school around the sixth grade. He sold newspapers and cigarettes to make money and eventually saved enough to open a bicycle shop. He later developed an interest in automobiles and opened one of the earliest automobile dealerships in the country. His avid interest in racing led him to create the Indianapolis Speedway in 1909. Fisher went on to make a fortune producing acetylene-lit lamps to allow driving at night through his Prest-O-Lite Corporation. In 1911, he sold his share of the Prest-O-Lite company for over $5 million to Union Carbide and turned his interest toward real estate developments in Miami and then Montauk. By the time he had purchased Montauk, Miami Beach was already a highly successful resort area catering to the elite, and Fisher envisioned the same for Montauk.

However, there was one hitch: the newly formed Long Island State Park Commission was also interested in acquiring land at Montauk to create a

An early photograph from Hither Hills State Park. *Long Island State Parks.*

public park and beach. Even though Fisher had an option to buy the portion of beachfront property that would become Hither Hills from the Benson family, the state prevailed. The park commission appropriated the land using a section of a New York State Conservation Law to back the claim that the state had the right to appropriate land for public use.

A public-use park was in direct conflict with Fisher's plan for an exclusive Montauk enclave; still, he went on to build roads, infrastructure and over thirty buildings in the downtown area. These include the present-day Montauk Manor, designed by Schultze and Weaver; the six-story Tower at Montauk, which once served as Fisher's offices and is now a condominium complex; the Montauk Yacht Club and Marina; and Montauk Downs Golf Club. However, only a fraction of his plan for Montauk was completed when the stock market crash of 1929 hit Carl Fisher hard. Many of his land holdings were put up for sale to compensate for his unpaid taxes, and his vision for a "Miami Beach in the North" was over.

Hither Hills State Park opened to the public in 1924. The name of the park comes from early English settlers who called the area the Hither, according to an account by Larry Penny for 27East. The "Hills" refer to the ridges and bluffs that were formed in this area during the end of the last ice age.

In 1978, the state acquired an additional 1,364 acres of bay and oceanfront land just to the west of Hither Hills on both the north and south side of Montauk Highway. Now called Napeague State Park, it is largely a preserve with no public facilities. Hither Hills State Park administers that property as well. Interestingly, the land at Napeague was never under threat from massive overdevelopment. In the late 1800s until the early 1900s, there were several bunker fish fertilizer factories there. The stench was quite overpowering and likely impeded any thought of major development in this area. The fish factories existed here until the 1960s, when the federal government passed restrictions on the harvesting of bunker, and the facilities were dismantled. Until recently, remnants of the large-scale Smith fish factory operations remained in the park along Napeague Bay. Today, there is only one building of the former factories remaining. It is located at the nearby Multi Aquaculture Systems, a fish farm that raises oysters, striped bass and other seafood and operates a small fish market off Cranberry Hole Road.

Napeague State Park is a passive-use facility. The north side of the park has hiking trails that pass through woodlands and wetlands, which may be why the Native Montauketts called the area Napeague, which translates to "land with water." The oceanside of the park is typically closed during the summer months, when endangered piping plover are breeding. There are no lifeguards, so no swimming is allowed. Napeague Beach abuts Hither Hills,

One of the former fish fertilizer factories in the area of Napeague State Park. *Katie Matejka.*

just to the east. There, you will find a bathhouse, camping facilities, picnic areas and stunning vistas overlooking the Atlantic Ocean, just like when it opened to the public in the 1920s. One of the hidden features of Hither Hills State Park is its thirty-foot-tall Walking Dunes, which continually shift position in the winds. Along the Walking Dunes loop on the north side of Hither Hills, east of Napeague Harbor, you might spot the Phantom Forest. This was once a healthy, thriving forest, but it is now slowly being uncovered by the shifting sands.

Hither Hills State Park and Napeague State Park remain largely untouched, except for campers and beachgoers who visit today. Nearby, in Montauk Village, you can see some of the buildings that would have been part of Carl Fisher's grand development scheme.

JONES BEACH STATE PARK

One of Long Island's most beloved parks, Jones Beach State Park in Wantagh, gets its name from Thomas Jones, an enigmatic character of Irish descent who once owned a large swath of land on the South Shore and may have been a pirate.

Jones's early life remains largely unknown, but it is widely believed that he fought under William III of England and for the deposed King James II. Historical records indicate that Jones was granted a license to operate as a privateer along the east coast of America in the 1690s. The distinction between a pirate and a privateer is often blurred, but the key difference is that pirates operate for their own gain, while privateers are authorized by governments to attack enemy ships during wartime. Privateers can be thought of as "official pirates." During this era, England commissioned privateers to attack Spanish and French ships to seize their treasures and resources for their cause.

Jones settled on Long Island after his father, John Jones, came to the New World and established himself in Flushing, New York. Thomas Jones married Freelove Townsend, the daughter of a wealthy merchant named Thomas Townsend, who had large landholdings in the Oyster Bay area and on the South Shore of Long Island, according to John Hanc, who wrote an extensive history of Jones Beach. As a wedding gift, Townsend gave the young couple three hundred acres of land in an area called Fort Neck, near the current Biltmore Shores in Massapequa, where Jones built a large brick house near Merrick Road and Massapequa Lake. Jones served in a local militia as a major before being appointed as a sheriff and then ranger general of Long Island, which was under British control at the time. He acquired an additional six thousand acres on the South Shore, where he established a prosperous whaling station along the barrier beach that would later bear his name. Jones's whaling station was located near the current state park beach. Jones died in 1713 and was buried in the cemetery of Old Grace Church in Massapequa, where you can still visit his and Freelove's tombstones.

In subsequent years, few people used Jones's beach, except for fishermen, as it was several miles from the mainland and navigation to its shores was challenging. Several United States Life-Saving Service (USLSS) stations were located in the area beginning in the mid-1800s. Over the years, many of the stations were destroyed by the elements, moved or consolidated. USLSS was the predecessor of the current United States Coast Guard. Station High Hill Beach was located just east of the former Zach's Inlet; Station Jones Beach was located on the east end of Jones Beach and was decommissioned in 1949; and Station Short Beach was located on the west end of Jones Beach and was demolished in 1990. That year, the coast guard consolidated operations at a newly built facility, renamed Station Jones Beach, on the west end of the Jones Beach barrier island that still operates today.

In the 1800s and 1900s, a summer cottage community called High Hill (named for the high sand dunes there) sprang up in the area around Zach's Bay, a mile east of the current water tower. Robert T. Willmarth built the Sportsmen's Hotel at High Hill Beach, and visitors arrived by ferry from Bellmore to enjoy fishing, swimming and other activities. A large open-air pavilion called Savage's Hotel and Casino provided music and dancing in the summer evenings. Other accommodations included the Breakers, which had a huge wraparound porch facing the ocean. There were dozens of buildings and cottages in the area, including a general store and post office. Cottage owners, who leased the land they built on, and other visitors to High Hill ventured to the area for relaxation and bathing.

According to Fred Schwab of the High Hill Striper Club, in his account "What About That Name…High Hill," the community had no telephones, electricity, gas or indoor plumbing. Summer residents drew water from wells. Residents cooked and lit their homes with kerosene. Savage's Pavilion was the center of the community, where beachgoers could find soda pop, ice cream, clam chowder, steamers and more. Another account in *Newsday Time Machine* describes how children were permitted in Savage's during the day, but it was strictly for adults at night, when liquor was served.

Watching water sports at High Hill Beach, near Savage's Hotel and Casino. *Wantagh Preservation Society.*

In the mid-1920s, when the Town of Oyster Bay and Town of Hempstead conveyed most of the barrier island to the Long Island State Park Commission to create its landmark Jones Beach State Park, several disputes arose. The heirs of a John Seaman came forward and claimed their family owned the land as part of a royal grant given in 1666. The case was carried all the way to the U.S. Supreme Court, but it was eventually determined that the claim did not stand and that the towns' conveyance to the state was upheld. Today, visitors pass over Seaman's Creek Bridge to Seaman's Island, where the Wantagh State Parkway continues to Jones Beach.

There were also questions about whether the cottages at High Hill Beach were in Oyster Bay or Hempstead. Eventually, it was determined they were located on land that was part of the Town of Hempstead's conveyance to the park commission. In the 1940s, some of the cottage owners who remained in what was now a public state park moved their houses by barge to the town of Babylon's West Gilgo Beach community, about five miles east. There are a few houses remaining on the north side of the park that were repurposed as parks maintenance buildings.

In addition, a group of sportsmen, including Solomon Guggenheim, sued the state after it took land they used on the east side of the park by eminent domain. According to Chester Blakelock in his 1957 description of

Looking south from Savage's Hotel at High Hill Beach. *Wantagh Preservation Society.*

Jones Beach State Park, the sportsmen used the land for their private club's recreational activities, including hunting and fishing. Their suit to retain ownership was denied, and for many years, the state maintained the land east of the park as a preserve. In the mid-1930s, the state turned it over to the Town of Oyster Bay. Today, the 550-acre site east of the state park is called the John F. Kennedy Memorial Wildlife Sanctuary. At its center is the freshwater Guggenheim Pond.

The construction of Jones Beach State Park was, to say the very least, quite an undertaking. Over forty million cubic yards of sand were dredged from South Oyster Bay to raise the beach fourteen feet to protect it from the unrelenting ocean surf. Previously, the area was made up of several small islands and marshland.

The park was opened in August 1929 and was an instant success. Visitors traveled on the newly built Wantagh Parkway, which, at that time, ran from Merrick Road to the beach. Just five years later, the state needed to build Meadowbrook Parkway to help alleviate the traffic. Transportation to Jones Beach State Park was done primarily via automobile along the scenic parkways. In the 1930s, the Bee Line Bus Company started running public buses to the park from the Wantagh and Freeport train stations, according to *Long Island Traditions*, which documents Long Island history and architecture. Visitors could enjoy the park for the price of a train and bus ticket.

As commissioner of the Long Island State Park Commission, Robert Moses's crowning achievement was Jones Beach State Park. Jones Beach State Park is one of the grandest public bathing facilities ever built in America. The bathhouses are a combination of Neo-Gothic and Art Deco styling, with some Moorish and medieval influences, while the grounds are of classic Beaux-Arts design. The buildings are made of Ohio sandstone and Barbizon brick, topped with limestone and copper accents. The landmark 1930s Italianate-style water tower, affectionately called the "pencil" or the "needle" by locals, is modeled after the fantastic campanile of St. Mark's Cathedral in Venice, Italy. It pumps fresh water to the park facilities.

Moses wanted to create a grand resort atmosphere, imagining the bathhouses, marine theater, restaurant and boardwalk would resemble a fabulous ocean liner. The boardwalk was built to resemble a ship's deck, with game areas offering shuffleboard and deck tennis and park workers dressed in navy-style uniforms. Walking along the boardwalk, trashcans are still disguised as horn-shaped shipboard vents. There are several nautical-theme tile mosaics throughout the park.

Jones Beach State Park was an instant success when it opened to the public in 1929. *Long Island State Parks.*

The Jones Beach Marine Theater, which originally opened in the early 1950s, primarily hosted musical extravaganzas, operettas and plays, many featuring well-known American band leader Guy Lombardo. You could also watch water ballet, synchronized swimming and other water-themed performances in the bay. In the early 1990s, the state added a second level to the original Marine Theater and its focus shifted to concerts under the promoter Ron Delsener. The theater is located on the west side of Zach's Bay, supposedly named for Zachariah James of Seaford, who owned the general store at High Hill Beach, according to an account by Birdsall James at the Wantagh Preservation Society.

Today, millions of people visit the 2,400-acre Jones Beach State Park to relax on its 6.5 miles of oceanfront beach or walk its boardwalk. It is located just 20 miles from New York City. There are food concession stands, restaurants, playgrounds, pools and a spray park. You can learn about the vast marine environment at the Jones Beach Energy and Nature Center. Anglers also come from all over to cast a line from the Field 10 fishing piers, where there is a fully equipped bait and tackle shop.

It truly is a park for the people.

MONTAUK POINT STATE PARK

Montauk Point State Park is located on the far-eastern tip of Long Island, almost 120 miles from New York City. Visitors and locals affectionately call it The End.

However, just as much as Montauk Point State Park is the final point east on Long Island's South Fork, it is also the first place in New York that boats traveling the shipping lanes just offshore pass as they head toward the Port of New York. The waters here have always been treacherous. Montauk is where the Atlantic Ocean meets the Block Island Sound, causing a visible race of converging tides. Many shipwrecks occur just offshore.

The most prominent feature that can be seen upon arrival at the park is the adjacent Montauk Point Lighthouse, built high on a bluff once called Turtle Hill, named for its resemblance to the shape of a turtle shell. The historic lighthouse was the first built in New York State, authorized by President George Washington in 1792. It was completed four years later to serve as a beacon and a warning about the dangerous waters, sandbars and hidden glacial boulders that lie just off Long Island's easternmost tip. John McComb, who also designed Gracie Mansion in New York City, which serves as the mayor's residence, was the architect for the lighthouse. McComb built Montauk Point Lighthouse using large Connecticut limestone blocks. Its tower stands 110 feet high, and its walls are 6 feet thick in parts to protect it from the howling winds and storms that are notorious in this part of Long Island. Sailors offshore have long seen its beacon from as far as 20 nautical miles away, a welcoming sight after a long transatlantic journey.

Montauk Point Lighthouse, built high on a bluff once called Turtle Hill. *Long Island State Parks.*

Civilian keepers maintained the lighthouse for over 140 years until the United States Coast Guard took over operations in 1939, according to the Montauk Historical Society. At that time, World War II was brewing in Europe. It is difficult not to notice a large, nearly five-story, boxy-looking white building standing directly next to the lighthouse tower. During World War II, this structure was equipped with radar and served as a surveillance tower to watch for a potential invasion by German vessels. In the event of a threat, the coast guard would have alerted nearby Camp Hero, which was equipped with heavy artillery and machine guns to defend the coast. Historically, the coast guard maintained two other stations in the area, one at Napeague and the other at Ditch Plains. These kept watch for ships in distress just off the rocky shore and provided search and rescue operations until the 1950s. In 1955, the stations were consolidated at a new facility at Star Island, near the inlet to Lake Montauk.

Recognizing the lighthouse's historical significance, the Montauk Historical Society began leasing the Montauk Lighthouse from the coast guard in 1987 and acquired it in 1996 to serve as a museum. It was named a National Historic Landmark in 2012 due to the significant role it has played in United States history. Inside, the museum includes displays on the rich and storied past of the lighthouse, its keepers, the United States Life-Saving Service and the shipwrecks that occurred just off the coast. Visitors can also climb the 137 steps to the top of the tower. There is a separate entry fee to the Montauk Point Lighthouse and Museum, and parking is located in the state park parking lot, which also charges a fee in season. While the lighthouse is inextricably linked to Montauk Point State Park, the state never owned it.

Montauk is named for the Montauketts, a group of Native Algonquians associated with the Pequots of Connecticut. Here, they fished and farmed and produced a form of currency called wampum made from local shells. There is some speculation that the name Montauk means "fort area," since the Montaukett had a fort at a nearby site now called Fort Hill. Often, settlers on Long Island called the Natives by the name they had for the land. There is evidence of fierce battles at Montauk near Fort Hill Cemetery, where the Montaukett set up defense against a raid by the Narragansetts in the 1650s. Wars and disease decimated a large portion of their population; however, there is evidence they coexisted with the early settlers. Remnants of their village are still located near Big Reed Pond.

Due to Montauk's abundant grassland, early English settlers used it as shared pastureland for grazing sheep and cattle. Deep Hollow Ranch, which is located near the entrance to the state park, is a remnant of that earlier time and is the oldest cattle ranch in the United States, established in 1658. The first English houses in Montauk were not built until the 1740s. The earliest house, once located near Hither Hills, has since burned down, but Second House still stands in Kirk Park on Fort Pond and Third House is located in Montauk County Park. Both were rebuilt after they also burned down, and they offer a glimpse into early settlement life. Third House was later used as a lodge to accommodate sportsmen and anglers who came to the area for its abundant fishing opportunities and game hunting. There were several wings added to it, along with cabins and other facilities.

While Montauk had sleepy beginnings, it twice came dangerously close to being heavily developed. In the 1870s, Arthur Benson purchased most of the Montauk Peninsula to use it as a sportsmen's retreat. Following his death, his heirs sold a portion of the land to Austin Corbin, a wealthy businessman from New York, who also owned the Long Island Rail Road. Corbin envisioned constructing a massive deep-water port where ships from overseas could offload their cargo as opposed to traveling the one hundred miles west to the Port of New York. From Montauk, the cargo would be loaded onto his railroad cars and transported to New York City. Facing several legal hurdles, he completed the rail to Montauk in 1895 but died shortly after, never realizing his dream. Today, visitors who pull into the Montauk Train Station can still see the six lines of unused railroad track Corbin built to transport containers and the deep-water pier at Fort Pond Bay, where large cargo ships would have docked. Montauk was nearly an industrialized port town for transatlantic cargo ships.

In the 1920s, another developer with big plans purchased a large swath of land from Benson's heirs, who still kept nine thousand acres themselves, including some of the present-day Montauk Point State Park. The developer Carl Fisher constructed several buildings west of the state park, including the current Montauk Manor, the six-story Tower at Montauk in town and the Montauk Yacht Club and Marina, along with roads and infrastructure as part of his vision to create a megaresort town resembling Miami Beach. While Fisher was busy developing his vision for an elite Montauk resort destination, Robert Moses at the Long Island State Park Commission had his own ideas. Moses was determined to create publicly accessible state parkland at Montauk, where all could enjoy the sparkling ocean beaches and wide-open spaces.

In 1924, the State Park Commission took 1,755 acres by eminent domain from the Benson family heirs, and that land became Hither Hills State Park. That same year, Moses seized 76 acres from the Benson estate and 10 acres from E.B. Mulford to create Montauk Point State Park. Additional acquisitions from Helen Hooper Brown, the Montauk Beach Development Corporation and Monell, Howell and Hodgkinson between 1925 and 1930 enlarged Montauk Point State Park to 335 acres. In the early 1960s, the state purchased another 528 acres, further increasing its size.

Meanwhile, Fisher kept up his pace of development. On the west side of Lake Montauk, he created the Montauk Golf and Racquet Club in 1927. Captain H.C.C. Tippet designed the original golf course, and famed New York City architect Stanford White built the original clubhouse. Stanford White, known for such public landmarks as the Washington Square Arch and the original Pennsylvania Station in New York City, built the clubhouse in a subdued Colonial Revival style with a large reception hall, dining facilities, restaurant, pro shop and lockers. The building later burned down.

After the stock market crash of 1929, Fisher lost much of his holdings. The golf course remained open but was run-down when a private investment group bought the Montauk Golf and Racquet Club in the 1960s. Robert Trent Jones Sr. redesigned the golf course, and it was opened as a private club

A historic aerial view of Montauk Downs State Park Golf Course, west of Montauk Point. The property was acquired by the state in 1978. *Montauk Library.*

in 1968. In 1978, New York State was considering constructing a golf course at nearby Hither Hills State Park but decided, instead, to purchase the golf course and rename it Montauk Downs State Park Golf Course. In 2003, the state commissioned Rees Jones, the son of Robert Trent Jones, to redesign the championship-length eighteen-hole course. There are also tennis courts, a driving range, a swimming pool and a full-service clubhouse.

Montauk Point State Park is located just a couple miles east of Montauk Downs State Park. Many of the beaches in the area, such as Ditch Plains, Hither Hills and the beaches just south of the main strip in downtown Montauk, feature beautiful white sand beaches and dunes. However, the shoreline at Montauk Point State Park is mostly rocky and rough, and there is no swimming permitted. Montauk Point State Park offers spectacular views of the Atlantic Ocean and the converging tides of the Block Island Sound. There are nature trails to hike, bike or cross-country ski on. In the winter months, visiting seals sun themselves on the rocks just offshore. The park is world-renowned for its surf fishing along the beach, especially when the striped bass are running. Experienced surfers also know the spot for its waves at Turtle Cove, just southwest of the lighthouse.

If Jones Beach State Park is the most iconic state park on the western side of Long Island, Montauk Point State Park is just as much a cultural landmark on the eastern point. Many travel here just to say they traveled to The End.

NISSEQUOGUE RIVER STATE PARK

The imposing abandoned buildings scattered throughout the grounds of Nissequogue River State Park in Kings Park provide striking evidence of the property's former use as a sprawling psychiatric facility that operated here for over one hundred years. Its scenic grounds and open vistas were initially envisioned to provide respite for the mentally ill who had endured abysmal conditions in the horribly overcrowded asylums of New York City.

Well before the psychiatric institution, Native people occupied the land in and around the Nissequogue River that runs alongside the park before it empties into the Long Island Sound. The Natives likely took advantage of the area's abundant fishing opportunities, like many visitors do today.

The name Nissequogue meant something like "land of clay and mud," which aptly describes the wetlands along the river. The current state park is situated on land that was part of the original deed given to Richard Smith by Lion Gardiner and Sachem Wyandanch in 1665. The Blydenburghs were one of the other early families who settled in this area around the 1700s, and they had considerable holdings throughout the township. In the 1850s, Noel Joseph Becar purchased a significant portion of the current park property from the estate of H.A. Blydenburgh.

Becar built a large, shingled house with a wraparound porch on the property north of St. Johnland Road, just east of the entrance to the northern section of the park. He was involved in the breeding of shorthorn bulls, and *The Wealth and Biography of the Wealthy Citizens of the City of New York* lists him as having in excess of $100,000, which would have meant he was

The original Becar mansion was used as the hospital superintendent's house after a renovation in 1900. *Kings Park Heritage Museum.*

quite well-to-do. The Becars would likely have been included in the activities of the upper-class society of the time. As evidence of this, Mrs. Noel Joseph Becar had her portrait drawn by famous Long Island artist William Sidney Mount. Her portrait, drawn in pencil on paper in 1861, is now part of the permanent collection of the Long Island Museum of Art, History and Carriages in Stony Brook, New York.

In the 1880s, the Kings County Lunatic Asylum was looking for a rural setting to ease overcrowding and poor conditions in the wards of New York City. The asylums in New York City were notorious for their poor treatment of the mentally ill due to overcrowding, insufficient funding and lack of knowledge about mental disease. Criminals, paupers and the mentally ill were typically confined in one place. The facility in Kings County, also called Brooklyn, was slightly more progressive than some other facilities. Its asylum in Flatbush housed only the poor and mentally ill, without the added stress of convicts, although conditions there were still awful and the asylum desperately needed more space.

In 1885, Kings County Asylum purchased 873 acres from the Becar estate and other landholders in the area, including Captain John H. Smith, S. Brush, S. Smith, J.B. Harned, John Sheridan, Charles Hallock, James Darling, John Kelly and Ebenezer Smith. The Becar estate was in disrepair and was no longer in use. The beautiful riverfront property was chosen as the ideal location for the Kings County Farm and Asylum. New theories were emerging at that time that said fresh air, good work and open space were beneficial for patients and could aid in their convalescence. Many advocated for country-like facilities to house patients. Originally called "lunatic farms," or, in more appropriate terms, "farm colonies," patients would participate in everyday activities and receive training in trades, farming, homemaking and self-sufficiency.

The first temporary buildings at Kings County Farm and Asylum housed only 55 patients. By 1889, 450 patients lived there, and it had its own dairy barn, piggery and slaughterhouse. Patients were engaged in working on the farm and would clear land for new buildings and help in the continued sustenance of the community. Four larger brick buildings were added in 1893. The original patient buildings were located near the former Becar mansion, which was still extant at that time. The Becar mansion was renovated in 1900 to house the superintendent of the hospital until a new brick superintendent's house, which still stands today, was constructed in the same vicinity.

Kings County Farm and Asylum was not the first colony in this small section of Long Island's North Shore. In 1866, Reverend William Augustus Muhlenberg, a minister and social innovator, had already established a small religious farm community just west of the current state park. Called

Early buildings from the state hospital at Kings Park. *Kings Park Heritage Museum.*

St. Johnland, Reverend Muhlenberg brought the aged, destitute, orphaned and impoverished to live in the colony based on principals of personal responsibility, morality, hard work and cooperation. At that time, the train stop in the hamlet was simply referred to as St. Johnland. In 1891, the Long Island Rail Road changed the name of the train station to Kings Park after the much larger hospital and its park-like setting, giving the hamlet its current name. The farm had its own power plant, water supply and a rail spur to bring in supplies. The rail spur, which crossed Main Street, is now a popular hike and bike trail.

In 1895, New York State took over operations of all previously county-owned facilities to create a more standardized model of care for the mentally ill. It renamed the Kings County facility Kings Park State Hospital. The state added many more buildings, including laundry facilities, a dairy and a fire department building. The state also opened a well-regarded school of nursing in 1906; it was registered with the New York State Department of Education and taught students how to care for patients until the mid-1970s.

The patient population continued to grow. In 1939, the Works Progress Administration (WPA) funded the construction of the twelve-story Building 93, designed by state architect William E. Haugaard. An unusual fact about the building: while it is twelve stories high on its right side (not counting the two-story attic space), it is only eleven stories on its left side. It is partially built into a small hill, so the first floor on the left side is a basement that once served as a recreation area. It is in that basement where it is believed a patient named Percy Crosby, who created the beloved *Skippy* cartoon character of the 1920s and 1930s, painted murals on the walls depicting patient life. Crosby died in the hospital in 1964, and the murals are now covered by graffiti.

The chronically ill and geriatric patients occupied the upper floors of the massive Building 93, which still stands at the site. Most of the patients slept dormitory style. Many of the ward buildings were constructed with wings to house women on one side and men on the other. Buildings contained dining halls, kitchens, staff offices, examination rooms and dayrooms. Patients used the dayrooms to play games, read, participate in crafts and hobbies, listen to music, talk and relax. Attendants generally adhered to a strict routine throughout the day, which included dressing, dining, socialization, occupational therapy and "lights out," which provided structure for them. Social activities were also created for the patients, including organized dances, plays, holiday events, games, activities and birthday celebrations. The

Recreational arts and crafts therapy for patients. *Kings Park Heritage Museum.*

hospital had its own theater, York Hall, for performances and ceremonies. Attendants worked with patients on everyday skills and hygiene and took them on walks and trips to town. Some of the patients were permitted to walk the scenic grounds alone.

A large influx of immigrants, many of them Irish, came to the area to live and work at the hospital as nurses, doctors, educators and attendants. Others worked in construction and maintenance. Some lived in houses and apartments on the property; others built houses in the small hamlet surrounding the hospital. The staff worked a rigorous schedule at the hospital; however, they received several days off per month. Around 1917, Joseph Brady operated a hotel called the Sound View that used to be located on the northern edge of the current park property. Staff would often spend time there on their days off, since many did not make enough money to travel far. The two-story hotel provided guests with rowboats to access nearby Short Beach, just across the river. In addition to hotel rooms, it had a restaurant, bar, parlor, card room and stairs down the bluff to the river. The state tore down the hotel in the 1920s, when it acquired the northernmost section of the property to create the veterans' hospital.

Kings Park State Hospital, later renamed Kings Park Psychiatric Center, reached a peak of 9,303 patients in 1954, which was more than the general population of the surrounding hamlet. Again, theories for the treatment of the mentally ill were evolving. At that time, there was less focus on occupational therapy, recreation and socialization. The practice of having patients work to support the farm colony ended after social reformists argued the state could not put patients to work without giving them fair wages. Doctors began using medications, such as Thorazine, Stelazine, Mellaril, Haldol and Elavil, to treat schizophrenia and depression, along with therapy and reintegration training to treat chemical imbalances of the brain. Into the late 1960s and 1970s, the atmosphere at the hospital changed. There was more of a sense of expediency to move patients out of the hospital and back into the community. The patient population plummeted throughout the 1980s and 1990s. In December 1996, Kings Park Psychiatric Center closed its doors, and the remaining patients were transferred to nearby Pilgrim State Psychiatric Center or to outpatient programs nearby. Two residential-style group homes were constructed at the entrance to integrate individuals into local neighborhoods.

The Kings Park property sat largely abandoned for several years. Any proposals or plans for massive redevelopment failed due to Kings Park community opposition and prohibitive clean-up costs. Many of the buildings are contaminated with asbestos and lead paint. There are demolished buildings buried on the site along with a vast network of underground tunnels, which made the thought of commercial development a challenge. In 2000, New York State Office of Parks, Recreation and Historic Preservation, which succeeded the Long Island State Park Commission, announced it would acquire 153 acres on the northernmost section of the property and create Nissequogue River State Park. In 2006, at the urging of the Kings Park community, local officials and preservationists, an additional 368 acres were transferred from the State Office of Mental Health to the New York State Office of Parks, Recreation and Historic Preservation. Over the next seventeen years, some of the buildings were removed, and the state installed playing fields, trails, a canoe/kayak ramp and a pavilion. A generous grant from the Charles and Helen Reichert Family Foundation, which funds local park initiatives, restored the former Veterans Administration building for use as park offices and a meeting space.

In 2023, the New York State Office of Parks, Recreation and Historic Preservation completed a master plan to outline the future of the park at the behest of the Nissequogue River State Park Foundation. In true democratic style, the state sought input from community members, park goers, conservationists and other stakeholders, who formed volunteer committees to provide direction and ideas. The plan includes reopening the old York Hall as a community theater and reusing the former laundry building on Old Dock Road as a farmers' market. There are plans to include a museum in at least one of the former buildings and create an arboretum in the area near the superintendent's house. The park is also a designated a State Bird Conservation Area. It features wetlands that are vital to the health of the Nissequogue River and Long Island Sound. The Long Island Greenbelt Trail runs through the property, and its trails offer scenic, bluff-top views of the waterfront.

While a handful of developers envisioned creating high-density housing and other intensive structures on the property over the years, the park will never be developed in that way. It will remain, as it had been in the hospital's early days, a place for those in need of respite. Out of respect for the thousands of patients who found their way here at the lowest point of their lives, one of the most poignant reminders of the park's history is a plaque near a patient burying ground that reads: "This plaque is placed in memory of the hundreds who are buried here. They lived with severe illness. May they rest in peace."

ORIENT BEACH STATE PARK

While the name Orient may seem like an unusual moniker for a place located in the Western Hemisphere, it is actually a geographic description of the far-eastern tip of Long Island's North Fork. The definition of *orient* is "a region lying to the east of a specified or implied point" or "the eastern region of the world."

The Native Algonquians, who inhabited the area until the first settlers arrived in the mid-1600s, previously called the land Poquatuck. The name Poquatuck means "open tidal river," according to the Oysterponds Historical Society (OHS). This name perhaps referenced the waters of the Narrow River and Long Beach Bay Tidal Wetlands area on the north side of the current state park.

The English Crown granted patents to the land in 1676, and early settlers used the land primarily for grazing cattle and sheep. The original patentees included the King, Terry, Vail, Young, Petty, Beebe, Rackett and Tuthill families. The settlers later used the rich and fertile land for growing a number of crops, including tobacco. During the Revolutionary War, the British were able to occupy the area due to its easy access and exposure from many points of entry. They freely helped themselves to cattle and other produce and used the site as a staging point to conduct raids on Patriot encampments in nearby Connecticut. Benedict Arnold served as a general here, with his base of operations in nearby Orient Village.

The early settlers in the area called it Oysterponds for the abundant oysters and other shellfish in the saltwater "ponds" separating East Marion from Orient. A wharf was constructed in 1740 in Orient Harbor, north of the

Orient Point Inn, 1930. *Southold Free Library, Whitaker Historical Collection.*

current state park, to provide a sheltered spot for commercial vessels as the small community continued to grow. The name Oysterponds was changed to Orient in 1836, after the United States Postal Service established a post office in the village. Apparently, Oysterponds was too similar to Oyster Bay, another town in western Long Island, according to OHS.

Orient was still a sleepy fishing and farming community when it started to attract sportsmen, anglers, tourists and beachgoers in the early 1800s. Some older visitors might remember an imposing structure that once stood on the north side of Route 25, just east of the entrance to Orient Beach State Park. It was a grand old hotel, four stories high, with a wraparound porch that offered sweeping vistas of the bay. Called the Orient Point Inn, the original building there dated to the 1670s, when it was originally a small house. During the Revolutionary War, the British took possession of the house and enlarged it by adding an adjacent garrison, according to Dr. George Cottrail in his 1959 *History of Orient.* After the war was over, the inn opened to the public in 1796. In 1834, Jonathan Fish Latham enlarged the inn to attract more tourists. A steamship from New York City arrived at its private dock facilities two times per week during the summer months to bring visitors to Orient to enjoy boating, fishing, swimming and other outdoor activities. The inn was also reputed to have excellent dining facilities and accommodations.

In addition to wealthy vacationers, the inn attracted politicians, actors, poets and writers. Some of its most famous guests included the poet Walt Whitman, who wrote portions of his well-known *Leaves of Grass* there. President Glover Cleveland was also reported to have been a guest at the inn, and James Fenimore Cooper, a writer known for his early American depictions, including *The Last of the Mohicans*, spent time in Orient. They likely fished in these waters and spent time on the rocky shoreline of Orient Beach. In the 1960s, the inn closed when it became economically unfeasible to comply with new fire safety codes. It sat abandoned and derelict, a shell of its former glory, until it was demolished in 1984. The docks that once brought throngs of visitors by steamship still exist just east of the park entrance.

Orient Beach State Park is located on a long peninsula that extends from Route 25 to Orient Beach and continues past the bathhouse and parking lot, where it juts out into Gardiner's Bay. At the far end of the peninsula is the Long Beach Bar "Bug" Lighthouse. Locals affectionately called the lighthouse Bug Light, because the original 1871 structure looked like a giant water bug when the rocks underneath it became immersed in water. The lighthouse is about an hour's walk along the beach from the main bathhouse. A fire destroyed the original Bug Light in 1963, and in 1989, Merlon Wiggins and several prominent Greenport families founded the East End Seaport Museum (EESM) and Marine Foundation to rebuild it. Today, EESM maintains the lighthouse and runs boat tours from nearby Greenport Village. The United States Coast Guard maintains the ten-inch solar-powered light, which still warns sailors of the dangerous sandbar in this area.

Much of the land along the Orient peninsula was owned by a co-op of sorts that dated to 1774, when the land was shared by the men of Orient Village for common use. Oversight of the area was conducted under the name Long Beach Association, and fees were collected for seasonal leases. Fishermen would set up shacks where they could harvest the abundant scallops and oysters. In 1862, Lewis A. Edwards leased a portion of the land to build the Atlantic Oil and Guano Company fertilizer factory, later called the Orient Guano Factory. Edwards, previously of Gardiner's Island, also served as a state senator. The legend on the trail map of Orient Beach State Park indicates a fish factory about two miles from the main parking lot along the unmarked trail. Aside from the fishing and fertilizer activity, the desolate and somewhat inhospitable terrain of Orient Beach State Park left the land largely untouched by development.

Long Beach Bar Lighthouse, affectionately called "Bug Light" because it resembled a water bug from a distance. *Southold Free Library, Whitaker Historical Collection.*

In 1929, the Village of Orient voted to dedicate the land to the Long Island State Park Commission for use as a public park. The King, Latham and North Sea Development Company dedicated additional acres. The state constructed a causeway along the peninsula to bring visitors to the beach area from Route 25. In 1947, after several storms and wash-overs, the state reconstructed the existing roadway on higher ground.

Today, the park has a bathhouse, playground and picnic facilities, along with forty-five thousand feet of waterfront on Gardiner's Bay, where visitors can swim, surf cast from the shore, kayak, windsurf and paddle board. There are also several hiking and biking trails. The park's rare maritime forest was designated a National Natural Landmark in 1980, and it contains red cedars, blackjack oak trees and prickly pear cactuses. The park is also an Audubon Important Bird Area, attracting bird enthusiasts looking for blue herons, egrets, black-crowned night herons, ospreys and other migratory birds.

On a side note, the area encompassing the state park and the nearby Cross Sound Ferry is typically referred to as Orient Point. The actual "point," or far-eastern tip of Long Island's North Fork, is not at the state park or at the ferry. It is located about a mile or so hike along a trail from Orient County Park, the site of the former Orient Point Inn.

PLANTING FIELDS ARBORETUM STATE HISTORIC PARK

It is difficult to decide which is more breathtaking at Planting Fields Arboretum State Historic Park in Oyster Bay: the ornate architecture and rich history of the fairy-tale mansion or the magical grounds and gardens filled with exotic flowering shrubs and towering trees.

The grand estate was once the home of William Robertson Coe and Mary Huttleson Rogers Coe. Coe made his fortune in insurance and railways and later became a sportsman, philanthropist and racehorse owner. Mrs. Coe was the heiress to a great fortune built by her father, Standard Oil magnate H.H. Rogers. Both were avid horticulturalists and named the property Planting Fields for the Indigenous Matinecocks who once lived in this area and used the land for planting crops.

Coe Hall is the second mansion to stand on the property. The first was an equally impressive brick and stucco Tudor-style mansion designed by Grosvenor Atterbury in 1906 for New York lawyer James Byrne and his wife, Helen. In 1913, Byrne sold the house and 350 acres of lawns and gardens skillfully designed by landscape architect James Greenleaf to the Coes. The Coe family acquired an additional 59 acres to complete the estate. Unfortunately, the original house burned to the ground only five years later when a workman's blowtorch accidentally ignited a ferocious blaze during a renovation. Valuable antiques and artwork filled the mansion, and the loss was valued at over $700,000 at the time.

The Coes commissioned famed architectural firm Walker and Gillette to rebuild the mansion that now stands on the site. The firm is known for their

Exterior view of Coe Hall. *Photograph by Mattie Edwards Hewitt, circa 1928; Planting Fields Foundation Archives.*

work on several Gilded Age mansions, including two former residences in Manhattan that now house the Italian and French Consulates. They also designed the Fuller building, Jacob Riis public housing and the East River Savings Bank in Manhattan, among many other works. When the new house was completed in the early 1920s, it evoked the feeling of a grand English estate. William Coe was, in fact, born in England, and the mansion closely resembles a sixteenth-century château called Moyns Park in Essex, England, that has a multi-gabled front façade and a cluster of massive chimneys very similar to Coe Hall. It was the height of the Gilded Age at that time, and hundreds of New York City's elite built extravagant residences along the North Shore of Long Island. Many of the wealthy families of that period sought to brand themselves as a sort of American royalty, and a lavish summer mansion on the Gold Coast of Long Island was a highly desirable status symbol. The Coes' main residence was located on the Upper East Side of Manhattan, and they also spent considerable time in Wyoming. Many of the Gilded Age country estates mimicked European castles and mansions. Their architects and designers would often travel to Italy, France or England to collect artifacts, furniture, windows, mantels, gates and sometimes even entire rooms from centuries-old estates to incorporate into their American commissions. Coe Hall is no exception.

The Coes had a unique vision for the interior of the home, desiring it to mimic a genuine European estate that would have been built gradually over the course of centuries. To achieve this look, Walker and Gillette used a combination of various styles from different eras to make it appear as though the house was centuries old. For example, the central hall has a medieval, Norman-style architecture that would have been typical of the eleventh and twelfth centuries, featuring Gothic beams, a flagstone floor and, most notably, a minstrels' gallery or balcony, where musical performers could play unobserved by guests.

In the dining room, the architects incorporated a sixteenth-century English Elizabethan style with impressively high windows and tapestries lining the wall. The stained-glass windows above a garden door in this room are of particular note. They are from Hever Castle in England, which was the birthplace of Anne Boleyn. Anne Boleyn was the second wife of King Henry VIII, and she held the title of queen of England from 1533 to 1536. Despite her efforts, she failed to conceive a male heir for Henry, who then ordered her execution on false charges. However, the daughter she bore, Elizabeth I, later became the queen of England. The windows depict the Boleyn family's coat of arms. The den in the house resembles a seventeenth-century English library or study. It features a heavily paneled interior with hidden bookshelves and seventeenth-century Dutch artwork. The mansion was built during Prohibition, when it was illegal to serve alcohol, so a secret panel in the room pops open to reveal a fully functioning hidden bar.

The Coe family also commissioned Charles of London, Everett Shinn, Robert Winthrop Chanler and Samuel Yellin to create the elegant atmosphere of the mansion interior. There are several noteworthy rooms to tour, including Mrs. Coe's bedroom, which has a fully restored mural on its walls and ceiling. The breakfast room contains a raised plaster mural of the Wyoming landscape by Robert Winthrop Chanler. It is one of only two of this type of Chanler work available for public viewing, according to the Planting Fields Foundation.

The grounds of the property are no less impressive. The original driveway to the house was located on Chicken Hill Road, where visitors would enter through a set of beautifully ornate eighteenth-century gates from a former English estate now called Carshalton Park in Sussex. The pedestals on either side of the gate are from the same estate. The original driveway meandered through stands of ornamental and native trees and flowering bushes and offered vistas of distant fields and gardens. Today, visitors can explore the beautiful gardens and paths on the grounds, which were designed by the world-renowned landscape architects the Olmsted brothers. They are known for their outstanding landscape designs at popular locations such

The Italian Garden with a reflecting pool, fountains, statues and a teahouse. *Author's collection.*

as New York City's Central Park, Yale University, the Biltmore Estate in North Carolina, Fort Greene Park and Grand Army Plaza in Brooklyn, the National Zoological Park in Washington, D.C., and the Rockefeller Estate at Kykuit in New York, among others. The firm also worked on the Bayard Cutting family estate at what is now Bayard Cutting Arboretum State Park on Long Island.

Close to the main house, the Italian Garden is perhaps one of the loveliest spots on the grounds. The area is sunken and contains a reflection pool with fountains and statues. Surrounding the pool, a decorative short wall contains ever-changing plantings and blooms. Mrs. Coe's bedroom window is located just above the garden, so she would have enjoyed the beautiful view. The teahouse, designed by Elsie de Wolfe, is located at the far end of the garden. It has an unusual curved shingled roof, beautifully painted murals and a trellised enclosure. Other exceptional areas to explore include the rose garden, the perennial garden, a sensory garden, a hydrangea garden, the dahlia garden and a dwarf conifer garden. There is also a cloister garden and children's playhouse.

The Coes were passionate about gardening, as is reflected in the grounds. The main greenhouse here was built by Lord and Burnham, which created the magnificent conservatory at the New York Botanical Garden. The greenhouse contains orchids, hibiscuses, palm trees, ferns and more. Another greenhouse on the property, the camellia greenhouse, contains over two hundred camellias that bloom at the height of Long Island winters, between January and March.

Sadly, Mrs. Mary Huttleson Rogers Coe died in 1924 at the age of forty-nine. She and William Robertson Coe and had four children together, William, Robert, Henry and Natalie. All went on to live highly successful lives as a railway magnate, diplomat, resort owner and Italian countess, respectively. Interestingly, many Coe descendants remained involved with the preservation of the estate and park through Planting Fields Foundation. In 1926, Mr. Coe married Caroline Graham Slaughter in a small ceremony in New York City. Slaughter was a divorcée and the granddaughter of a Texas governor. In 1948, Mr. Coe arranged for the mansion and garden to be gifted to New York State to be used as a school of horticulture following his death, according to Planting Fields Foundation. After Mr. Coe died of an asthma attack in 1955 at the age of eighty-five, State University of New York (SUNY) operated a college there from 1957 to 1962. Called the State University Center on Long Island at Oyster Bay–Agricultural and Technical Institute, the school, in its first year, welcomed 144 students, who were housed in converted horse stables. The mansion on the university campus was nicknamed Coe Hall by its students, and the name has since stuck.

Mrs. Caroline Graham Slaughter Coe continued to live in the cottage on the property, surrounded by six acres of land, until she died there of a heart attack in 1960. SUNY maintained the property until 1971, when it transferred the estate to Long Island State Parks. At that time, New York State was facing a massive debt crisis, and little funding was available to maintain the estate. Fortunately, before his death, Mr. Coe had created the Planting Fields Foundation with an endowment. To this day, the foundation maintains and interprets the property for visitors and has been a major component in the continued preservation of this grand country estate.

The foundation provides educational opportunities for school groups and master gardeners and hosts many public cultural events, as well as lectures, exhibitions, concerts, performances and family events. It works in partnership with the New York State Office of Parks, Recreation and Historic Preservation to serve as a steward of this cultural gem.

ROBERT MOSES STATE PARK

Robert Moses State Park, located in Babylon, was the first New York State Park on Long Island. In 1893, the state acquired two hundred acres of stunning oceanfront along the western tip of Fire Island's barrier beach and opened it to the public in 1908. Initially, it was named Fire Island State Park.

At the time of the park's opening, there were no bridges or roads leading to it, nor were there any other state parks or parkways yet constructed on Long Island, for that matter. These would all come later when Governor Alfred Smith appointed Robert Moses to head up the newly formed Long Island State Park Commission in 1924. Moses embarked on an unprecedented mission to acquire vast acres across the state to create public parkland. As a result, the park was later renamed Robert Moses State Park in honor of his work.

Born in 1888 in New Haven, Connecticut, Moses grew up in midtown Manhattan near Fifth Avenue. His father was a department store owner, and his mother worked on philanthropic pursuits in the Jewish community on the Lower East Side. Moses received a doctorate degree in political science from Columbia University in 1914. He became interested in civil works and reforms early in his career, starting at the New York City Bureau of Municipal Research. As he worked his way up, he held many positions and titles and became a close ally of Governor Alfred Smith. The two would go on to develop some of the most ambitious public works projects of the time, including bridges, highways, two World's Fairs in Queens, massive state parks and hundreds of playgrounds across the state.

Robert Moses (*center*) enjoyed the beach and swimming on the South Shore. *Long Island State Parks*.

Robert Moses served as the head of the Long Island State Park Commission from its formation in 1924 until 1963. During his time as Long Island State Park commissioner, he launched an ambitious campaign to create an interconnected system of parks and parkways to transport people from the sweltering heat of Manhattan to Long Island's scenic and recreational outdoor opportunities. Robert Moses's grand vision was to create parks and parkways for the public's health and enjoyment. On Long Island, Moses acted with the support of Governor Alfred Smith, and when he retired, New York State had over two and a half million acres of state park land. On Long Island, this included state parks, numerous preserves, historic sites and golf courses, along with miles of tree-lined state parkways that serve as a peaceful contrast to other heavily trafficked commercial roadways. An interesting note about the miles of roadways and bridges Moses so lavishly constructed to bring people to the parks: Moses himself never learned to drive.

After retiring from public service, Moses chose to live in Babylon, close to the causeway and the state park, both of which bear his name today.

He was fond of spending his days at the beach and swimming in the ocean waters along the South Shore's barrier islands. He resided on Thompson Avenue in Babylon and rented a beach house in nearby Oak Beach and, later, Gilgo Beach. In 1981, at the age of ninety-two, he suffered heart failure and was taken to Good Samaritan Hospital, which is still located on the west side of the bridge Moses built across the Great South Bay. From his hospital window, he most likely had a full view of the Robert Moses Causeway and Bridge.

Robert Moses State Park is located on the western tip of Fire Island. The Native Secatogues in the Islip area called it Seal Island for the large number of wintering seals there, according to Lee E. Koppelman and Seth Forman in their definitive book on Fire Island. In the 1650s, Isaac Stratford constructed a whaling station here and called it Whalehouse Point. It continued as an important whaling site in the 1700s and 1800s, when there were also rumors of pirates in the area, like Captain Kidd, who allegedly buried plundered treasure there in the late 1600s. According to Islip town historian George Munkenbeck, new research shows Kidd might have spent time in the Islip area before he was sent back to England and executed for his seafaring crimes.

Nobody is completely sure why the early settlers called it Fire Island. It could be a misspelling of Five Islands or a misinterpretation of Vier Island, or "Four Islands," on an early Dutch map. The number of barrier islands running from Long Beach to Hampton Bays on Long Island's South Shore changed over the years as various inlets and breeches changed the landscape.

Visitors have enjoyed the beach at Fire Island since before the state park was established. In the 1850s, David Sammis built the Surf Hotel and guest cottages just east of the Fire Island Lighthouse. The hotel was accessible only by ferry or private boat, so Sammis built a dock to accommodate the boats arriving from Babylon. Wealthy patrons from New York City flocked to the Surf Hotel, which could accommodate five hundred guests and was the largest hotel ever built on Fire Island. There, the elite of the time enjoyed the white sand beaches, the healthful air, sailing and fishing until 1892, when there was a cholera scare aboard an immigrant ship, the SS *Normannia*, that approached New York Harbor. State Health Department officials scrambled to find a place to quarantine the passengers. The remote Surf Hotel seemed like the ideal spot, so passengers aboard the ship were taken there to ease the emergency. Cholera was a deadly disease at the time, and residents in the area were frightened. They actually armed themselves, took possession of the dock and threatened to burn down the hotel to prevent arrivals. After the crisis passed, the hotel eventually reopened to the public. However, it never regained its former popularity and burned down in 1918.

The federal government built the first Fire Island Lighthouse, located just east of Robert Moses State Park, in 1825. However, at only 75 feet tall, that tower was not tall enough to fully aid in navigation. You can still see the foundation of the original tower near the current lighthouse. Built in 1858, the black-and-white lighthouse that stands today is 168 feet tall. It is the tallest lighthouse on Long Island, and visitors can climb to the top and explore the nautical museum. However, it is not part of the state park. The Fire Island Lighthouse Preservation Society operates the museum. That group worked to ensure the preservation of the lighthouse after the government decommissioned it in 1974. The museum contains impressive exhibits and interactive displays about the lighthouse and its role in guiding ships just offshore.

To the west of the lighthouse, the federal government constructed United States Life-Saving Service (USLSS) no. 25 Station Fire Island in 1849. This station was one of many along the coast of Long Island that provided search and rescue services in the event of a shipwreck. According to the Fire Island Preservation Society, there were numerous shipwrecks that occurred here between 1849 and 1914. The brave members of the life-saving station responded to these emergencies regardless of the weather, in the summer and winter, in order to rescue stranded passengers. Sometimes, they would use a surfboat to reach the affected vessel, while at other times, they set up a life-saving apparatus on the beach that used a hawser line to connect them to the ship in distress. They sent a harness known as a breeches buoy across the line, above the pounding surf, to haul passengers ashore one by one. This was extremely treacherous work. The original USLSS no. 25 moved several times due to storm surges or wash-overs. Originally, the station was located near the intersection where the wooden boardwalk from Robert Moses Field 5 crosses over the dirt road to the lighthouse.

In 1915, the United States Life-Saving Service was renamed the United States Coast Guard, and all remaining stations were renumbered. In 1936, Coast Guard Station Fire Island no. 83 was moved near its current location on the west end of Fire Island on the bay side. The station suffered extensive damage during the hurricane of 1938, when a storm surge battered Fire Island. Today, Coast Guard Station no. 83 has been expanded, and its signature red roof can be seen just east of the bridge that leads into the park.

In 1924, the U.S. government transferred an additional six hundred acres west of the lighthouse to the State of New York to enlarge the original two-

The former U.S. Life-Saving Service Station–Fire Island with the lighthouse tower behind it. *Fire Island Preservation Society.*

hundred-acre Fire Island State Park, although it stipulated that the coast guard retained the right to use any portion of the land, as it needed, to perform its functions at any time.

During these early days, reaching Fire Island State Park was possible only through private boats or intermittent ferries from Babylon or Bayshore, which took more than an hour. In 1930, the Long Island State Park Commission acquired both Captree State Park and the land for Ocean Parkway with a vision of creating a stretch of oceanfront parks and parkways from New York City to Montauk. A portion of the Ocean Parkway was constructed between Jones Beach State Park and Captree State Park. The planned parkway would have continued across a bridge to Fire Island, traveling through its communities and wilderness, and then looped north again by bridge to the mainland near Moriches. Drivers could then continue east to Montauk, creating a nearly one-hundred-mile-long stretch of mostly oceanfront parkway. The ambitious plan to construct a parkway the full length of Fire Island was abandoned due to heavy opposition, and today, there are no roads east of the park.

In 1954, the park commission built a causeway from Southern State Parkway to Babylon, along with a two-mile-long bridge and a drawbridge across the State Boat Channel to Captree State Park. From there, visitors could take hourly ferries to the ocean beaches at Fire Island State Park in just fifteen minutes. The ferries ran until the final span connecting Captree Island to Fire Island opened in 1964, completing the link from the mainland. In the same year, the causeway and park were named after Robert Moses, a man whose July 30, 1981 obituary in the *New York Times* described him as being, "in every sense of the word, New York's master builder. Neither an architect, a planner, a lawyer nor even, in the strictest sense, a politician, he changed the face of the state more than anyone who was."

In 1968, a sister bridge connecting the mainland to Captree was completed to ease traffic. The second bridge has three lanes, allowing visitors to cross the boat channel and the Great South Bay departing the park, while the original bridge, now leading into the park, has two lanes. That year, the state also completed the two-hundred-foot-high ornate water tower in the center of the traffic circle, which pumps fresh water into the park. The tall brick tower may look similar to Jones Beach's Italianate tower, but it has its own characteristic cylindrical design and pointed peak. Long Island children have affectionately called it the pencil over the decades.

From the traffic circle, visitors can proceed to the public Fields 2 and 3 to the west and Fields 4 and 5 to the east. After the great hurricane of 1938 devastated most of the park, the state built a white wooden bathhouse two miles west of the lighthouse. That bathhouse at Field 3 is the oldest bathhouse at the park. The state built the other three brick bathhouses later. Each field has a concession stand, beach shop, restrooms and outdoor showers. Scenic picnic areas with grills and tables are located at Fields 2,

3 and 4. Overall, there are five miles of ocean beach where visitors can swim, surf and surf cast. The park also has an eighteen-hole pitch and putt golf course at Field 2, and the original ferry marina and fishing piers are accessible via an underpass at Field 3.

Field 5 has a large playground for children and a boardwalk that runs along the ocean. It is also the public entrance to the rest of Fire Island. A three-quarter-mile-long boardwalk through the dunes leads to Fire Island Lighthouse. To the east are the communities of Fire Island, including Kismet, Ocean Beach, Cherry Grove, Fire Island Pines and Davis Park, scattered amid the spectacular Fire Island National Seashore. These communities predate the federally designated Fire Island National Seashore, which is otherwise undeveloped for thirty miles along the shore. There are no roads except the dirt Burma Road, and points east are accessible only by ferry, foot or bicycle.

SHADMOOR STATE PARK

Shadmoor State Park, located in Montauk, is a small and often overlooked state park. Tourists visiting Montauk usually prefer to visit the larger and more popular state parks, such as Montauk Point State Park and Hither Hills State Park.

The name Shadmoor comes from the combination of the area's two main features: the high moors along the bluff-top preserve and the shadbush, or juneberry, that fills the grounds. It may be one of the smallest state parks; however, Shadmoor played a role in United States history in the Spanish-American War and World War II. Furthermore, its towering, fluted bluffs are highly unusual for the South Shore of Long Island, which consists mostly of low-lying white sand beaches.

Shadmoor, like much of the rest of Montauk, was once home to the Indigenous Montaukett. The name Montauk may mean something like "fort area" in the Algonquin language, and it refers to both the land and the people who lived in the area before early settlers. The stretch of land from Hither Hills and Napeague is quite low and flat; however, the land starts to rise noticeably where Shadmoor State Park is located. The Natives called it Nominick, which means "land that can be seen from far away," and indeed, it can be.

The Montauk peninsula is largely composed of grassland that was excellent for grazing cattle and sheep. The early settlers leased the land from the Native people and used it as pasture prior to any significant development. While nearby settlers in East Hampton built their houses in the mid-1600s,

Montauk would not have any permanent residents until the first house was built in the 1740s, and even then, it was only keepers who tended the shared pasture and animals who occupied it.

In the mid- to late 1800s, some visitors found their way to Montauk to take advantage of the area's recreation and sporting opportunities. However, parts of Montauk were still largely isolated. For this reason, it served as the ideal camp for soldiers returning from Cuba during the Spanish-American War in 1898, as they needed to quarantine due to an outbreak of yellow fever. The quarantine ships arrived at Fort Pond Bay, and the massive four-thousand-acre Camp Wikoff spanned the area in the north between the bay and Lake Montauk and down to Ditch Plains and Shadmoor State Park. At one point, there were nearly thirty thousand men in quarantine there. It was named Camp Wikoff after Colonel Charles A. Wikoff of the Twenty-Second U.S. Infantry, who was killed during the Battle of San Juan Hill in Santiago de Cuba. The Red Cross provided assistance to care for the men, and the camp contained thousands of tents, operating rooms and dining facilities. Colonel Theodore Roosevelt, who went on to become the twenty-sixth president of the United States, and his Rough Riders were some of the camp's better-known guests. When they arrived at Fort Pond Bay, cheers awaited them for their bravery during battle.

At that time, the nearby United States Life-Saving Service (USLSS) Station no. 7 Ditch Plain also patrolled the beachfront of Shadmoor State Park. The station was constructed in 1855 and was later renamed Coast Guard Station no. 65. In the late 1800s, the USLSS had stations approximately every three miles along the South Shore, and the brave men stationed there performed life-saving operations for the numerous shipwrecks that occurred off the rocky coastline.

Coast Guard Station no. 65 and the grounds of Shadmoor State Park were also active during World War II. At that time, the need for protecting the coastline became imperative, as hostile German submarines made their way into the East Coast shipping lanes just offshore. The Northeast Defense Command was created in 1941 to prepare for and defend the East Coast against potential enemy invasion. It later became the Eastern Defense Command and was in charge of army coastal defense and antiaircraft operations. Montauk became a key strategic component in the surveillance and defense operations of the United States in the event of infiltration or attack. Shadmoor State Park was fitted with wood-encased concrete bunkers that once held heavy artillery as part of defense efforts, according to New York State Parks. Today, you can still see remnants of the bunkers

A World War II bunker still extant at Shadmoor State Park. *Katie Matejka.*

along Bunker Lane, one of the trails in the park. The nearby station was decommissioned in the 1950s.

In 2000, New York State purchased the land from private investors who had intended to subdivide it for housing. The acquisition was a combined effort with Suffolk County, the Town of East Hampton and the Nature Conservancy. The combined entities paid $17.7 million to save Shadmoor from development after a decades-long battle and community activism by the Concerned Citizens of Montauk, along with the Seaside Avenue Association, Surfside Estates Association and other entities. The groups argued against development, stating, among other points, that the distinctive flora and unusual geological features are unique along the entire Eastern Seaboard and therefore should be preserved.

In her book *Holding Back the Tide: The Thirty-Five Year Struggle to Save Montauk*, Joan Powers Porco describes the passion these groups had for this relatively small park. Shadmoor is home to an extremely rare plant called sandplain gerardia, which likely once covered much of the Montauk coastline. The area's other plants include sedge, winterberry holly and black cherry. The shadbushes on this high moorland rarely grow higher than a few feet due to the punishing winds coming off the Atlantic Ocean. The vegetation is stunted yet well adapted to the environment.

The park ends at a high bluff that was formed during the last ice age, when glaciers moved and shaped the land that became Long Island. The cliff is composed primarily of clay, and the numerous streams that run through the park carve unusual channels where the precipice meets the beach below. Between some of the channels are hoodoo-like formations, which are fragile, pillar-like sculptures that constantly shift and change with the wind and tides. Montauk is the only place on the South Shore where high bluffs meet the Atlantic Ocean.

At only 99 acres, Shadmoor State Park is one of the smallest state parks on Long Island. It has 2,400 feet of oceanfront beach accessible via a staircase down the cliff. Visitors come to picnic, birdwatch and fish. Swimming is not permitted. While its shoreline may not seems as aesthetically pleasing as the miles of smooth white sand beaches where sunbathers flock in the summertime, Shadmoor is captivating in its own distinct way.

SUNKEN MEADOW/GOVERNOR ALFRED E. SMITH STATE PARK

Sunken Meadow State Park gets its name from a vast, sundrenched meadow that once extended from the calm waters of the Long Island Sound to the high, forested bluffs behind the park. While parking lots have replaced much of the sprawling lowland meadow, it is still apparent how the area would have appeared sunken from atop the waterfront cliffs.

The far-eastern edge of the park is located near the mouth of the Nissequogue River, which may have meant something like "land of clay and mud" to the Native people who once lived in this part of Long Island. That part of the park still contains environmentally significant tidal flats bisected by Sunken Meadow Creek and bordered by high bluffs.

The centerpiece of the park is its beautiful sandy beach, three-quarter-mile-long boardwalk and shady picnic grounds in the fields behind the beach. Even before the Long Island State Park Commission began purchasing several parcels of land to create the park in 1926, locals in the area used it as a beach. The original entrance was located on Sunken Meadow Road just across from what is now Kohr Road. Tyler Avenue, no longer in existence, ran down to Sunken Meadow Creek, and visitors could cross along a quarter-mile-long bridge.

One of the earliest acquisitions the Long Island State Park Commission made to create a portion of Sunken Meadow State Park came from the estate of Antoinette Storrs Lamb. Antoinette's husband, George B. Lamb, had purchased a beautiful circa-1754 Georgian-style mansion and two thousand acres of land along the Sunken Meadow Creek from Caroline

A view of the "meadow" from Gus Kohr's North Shore Hotel, near the former entrance on Sunken Meadow Road. *Kings Park Heritage Museum.*

Child Platt in 1903. Caroline Platt was a descendent of Zephaniah Platt, whose father, Jonas Platt, acquired the land from Richard Smith II in 1717. The Smiths were the original patentees of the land that comprises the Town of Smithtown and the hamlet of Kings Park, where Sunken Meadow is located. The historic Platt mansion still stands on Sunken Meadow Road and has an historic marker commemorating the role the property and Zephaniah Platt played during the Revolutionary War.

Zephaniah Platt was an ardent Patriot during the British occupation of Long Island from 1777 to 1784. Many Long Islanders despised the British presence here, and along the North Shore, they formed a brave network of spies who would gather information about British troop movements. Zephaniah Platt allowed the Patriots to hide their whaleboats in a barn on his property. They used these boats covertly to row across the Long Island Sound to Connecticut to bring intelligence about British troop movement to couriers who would take the information directly to General George Washington. When the British, who had a nearby encampment named Fort Slongo, discovered the operation, they apprehended Zephaniah Platt and sent him to one of the notorious British prison ships in New York Harbor. Prisoners lived in deplorable conditions aboard these filthy detention ships and often succumbed to a miserable death. Since Zephaniah was an older man at the time, his daughter Dorothea pleaded for his release, which the British granted. But Zephaniah died a short time later in his home on Sunken Meadow Creek, as he never fully recovered from his harsh confinement.

For nearly two centuries, the Platt family occupied the house and property before they sold it to the Lambs, who used it as a hunting preserve. The state purchased 248 acres of land from the Lambs between 1926 and 1929. The state began cobbling together other parcels to create the park. In 1928, the Town of Smithtown deeded an initial 400 feet of waterfront to the state; George E. Conklin sold to the state a .41-acre lot and a house; and the Fort Salonga Beach Corporation dedicated a .38-acre parcel of wooded uplands.

In the same year, the state bought sixty-six acres of forested uplands and sound frontage from the Society of St. Johnland. This society has an interesting history. It was established in 1866 by Reverend William Augustus Muhlenberg, who had a vision of creating a farming and industrial settlement on the picturesque bluffs of the Long Island Sound. This idea was born after he witnessed the miserable conditions of the poor and destitute in New York City. He raised funds to purchase a four-hundred-acre farm on the bluffs adjacent to the future park, where he could provide respite to society's most vulnerable. He enlarged the original Smith farmhouse and added several more houses on the farm to house orphans and pensioners. The community and farm, guided by Christian values, operated for eighty years as an orphanage, providing care for young and elderly residents. However, the management eventually realized that providing adequate care for both age groups would require different types of care staff. Consequently, the orphanage was discontinued in the 1950s. Today, St. Johnland still exists in Kings Park next to Sunken Meadow. It now operates as a modern retirement home and nursing facility, providing excellent care to the elderly.

In 1930, the Long Island State Park Commission appropriated additional properties that were owned by Antoinette Storrs Lamb, Carll S. Burr and Irving and Kate Vause. The Vause house, which was originally situated in the lower meadow on Sunken Meadow Creek, had already been relocated by barge in 1928; it was moved down the creek to its current position on the east side of the Kings Park Bluff. The Vause house still stands there at the end of Old Dock Road today.

The Vause house was not the only house located at Sunken Meadow that was moved when the state purchased property. Moving houses back then was much more common than it is today. Houses were generally well-built back then, and it was sometimes less work to move them than it was to build a new one.

Young residents of the Society of St. Johnland, a community that cared for orphans and pensioners, 1935. *Kings Park Heritage Museum.*

Gus Kohr's North Shore Hotel stood at the old local entrance to the park on Sunken Meadow Road, near Kohr Road. Built around 1913, it accommodated several notable visitors, as indicated by the society pages of the *Long Islander* newspaper. The hotel allowed easy access to the beach and had a stunning view of the meadow and the waters of the sound. When the state acquired the hotel, it cut it in half. One side of it remains on state park property and is the park superintendent's house. The other piece of the hotel was moved farther down Sunken Meadow Road, east of Kohr Road, and is now used as a private residence.

There were several other early houses in the area of the current parkway bridge and east of it, including property owned by B. Blydenburg and L. Hallock, according to a 1917 Belcher Hyde map from the Kings Park Heritage Museum. In 1930, H.D. Franciola's company Mascengo Inc. sold to the state a 13.6-acre parcel along with a twenty-five-room Colonial mansion on the creek that had previously been owned by S. Leroy Ackerly. According to a 1930 *New York Times* article, the state remodeled the Franciola house for use as a public tearoom and restaurant; however, it is unclear if it ever operated as such. It was located near parking field 4 on the south side of the creek and was knocked down in 1955.

The North Shore Hotel was spilt in half. One half is currently used as the superintendent's house at the park. *Kings Park Heritage Museum.*

The original state park opened in 1929 on 520 acres, and the state continued acquiring properties until the beachfront park comprised an eventual 1,286 acres. Sunken Meadow Creek was drastically rerouted and dammed to expand the parking fields and widen the beach on the east side of the park. Superstorm Sandy destroyed the old dam in 2012, and the state decided that it would be healthier for the wetlands there not to rebuild it. A footbridge now stands there, linking the meadow to the bluffs.

In order to expand the park, additional land was acquired over time. One such acquisition was the 179-acre Daybreak Estate, previously owned by John Shields. The Shields mansion, built in 1916, was considered a luxurious residence among the wealthy in the small community enclave of Fort Salonga. It was situated just 2,000 feet away from the cliff's edge, overlooking the Long Island Sound. A real estate developer had planned to subdivide the property for housing in 1958, but the state intervened and purchased it to expand Sunken Meadow instead. The mansion was located west of the bathhouse at Field 2, and the road leading to it was called Naples Avenue. Today, that road leads to the golf course at the park, where there were also several more houses moved, demolished or reused to make way for the links.

The Sunken Meadow golf course, designed by golf course architect Alfred H. Tull, was completed in 1964. Tull is noteworthy for incorporating the layout of his courses directly on the existing topographical features of the land without altering the land in any significant way. The course at Sunken Meadow is noteworthy for its surrounding woodlands and hilly terrain, and the view of the Long Island Sound from the golf driving range is spectacular.

The most recent addition to the park came in 2002, when Robert and Roberta Gouldstone of England sold twenty-two acres of wetlands and woods to the state for $1.25 million. The property, also on the west side near the golf course, includes an amazing stand of white cedar trees and a gravestone from one of the earliest farming families in the area, the Treadwells. This portion also contains streams and ponds, as well as a fish ladder, and has largely been left untouched.

In 1992, Sunken Meadow was renamed Governor Alfred E. Smith Sunken Meadow State Park. While Robert Moses receives much of the recognition and credit for creating the state parks and parkways on Long Island, Governor Smith was the one who gave him the ability to do so. Smith, known for his progressive agenda in education, workplace rights, housing and welfare was an adamant supporter of creating parks for the public's use and enjoyment. He served four terms as governor of New York between 1918 and 1928. In

1924, Smith sponsored the legislation to create the initial State Council of Parks and the regional park commissions. Voters approved an unprecedented $15 million bond to acquire new public parkland. Over the course of just one decade, fifty-five new parks were established throughout the state.

Sunken Meadow was one of the early state park acquisitions, although improvements and additions were made over the course of several decades. The state conducted major improvements at the park between 2014 and 2019, when the main bathhouse and tolls were refurbished and a number of environmental projects were undertaken to improve the park's ecology.

Today, the park offers a wide range of landscapes and scenery. From the high, wooded bluffs and picnic grounds among the fields and trees, to the three-quarter-mile-long boardwalk along the expansive shoreline, it also contains valuable wetlands that are home to a variety of wildlife. Visitors can enjoy the hiking and biking trails, kayaking, barbecuing, swimming in the sound and playing a round of golf. Just as the park was compiled through the acquisition of many different parcels of land, it offers a variety of different experiences to be enjoyed by the many visitors who come here.

25

TRAIL VIEW STATE PARK

Trail View State Park is one of the more unusual state parks on Long Island. It is only a quarter-mile wide at its widest point, but it stretches for seven miles from Cold Spring Harbor State Park in the north to Bethpage State Park in the south.

With such an irregular shape, it is easy to envision that the Long Island State Park Commission acquired this long, narrow piece of greenspace to create a parkway. In 1961, Park Commissioner Robert Moses announced plans to build a spur from Bethpage State Park that would run north, all the way to the recently acquired Caumsett State Park in the Lloyd Harbor area of Huntington.

Its intended name was Caumsett State Parkway, and it would have connected with Southern State Parkway, Northern State Parkway and the Long Island Expressway. Travelers along the Long Island Expressway may notice today that there is no exit 47, which would have been the exit to the never-built parkway.

Prior to the park commission's acquisition, the land along the route was sparsely populated and was composed mostly of farms and woodland. Moses's plan was to turn the newly acquired Caumsett State Park, once the former Marshall Field III estate, into a developed park with areas for swimming, golf, recreation and boating. Millions of visitors were expected to visit the park once access was completed. The parkway was to run through scenic and historic Cold Spring Harbor and continue up Lloyd's Neck, where it would have transplanted many homes. In fact, a

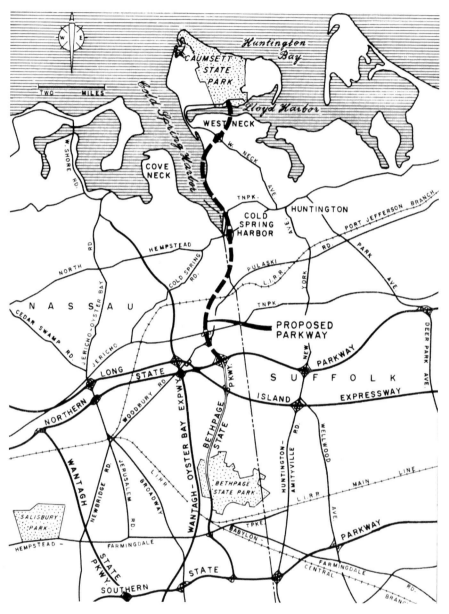

The dotted line shows the planned Caumsett State Parkway, now Trail View State Park. *Huntington Town Historians Office.*

small neighborhood where the current Cold Spring Harbor Library is now located was removed in preparation for the parkway. At that time, Terrace Drive was connected to a road called Harbor View, where the houses were located.

The residents of the North Shore promised a fight. Groups such as the Huntington Taxpayers Committee and Cold Spring Harbor Civic Association balked at the prospect of a parkway running through their community and a Caumsett State Park that would have been akin to the intense usage at Sunken Meadow State Park or Heckscher State Park.

The park commission was facing its own challenges. A deepening recession and fiscal crisis in the 1960s and 1970s and a subsequent lack of available funding brought the plans to a sobering halt. New York State was heavily in debt and facing massive budget gaps following an era of broad-scale public spending. Despite acquiring the land for the parkway, the state shelved its plans for Caumsett State Park and Caumsett State Parkway and the land for the parkway sat untouched.

In 2002, New York State agreed to dedicate the right of way for the Caumsett State Parkway as a state park at the urging of environmentalists, nearby residents and, most notably, the Long Island Greenbelt Trail Conference. The Greenbelt Trail Conference recognized the potential of the long, narrow greenspace as early as the 1980s. That group is a unique grassroots organization dedicated to preserving open space, and today, its members maintain more than two hundred miles of hiking trails on Long Island. The group had already begun forging a trail through the seven-and-a-half-mile-long Trail View property as part of a larger plan to create a twenty-mile-long Nassau–Suffolk Greenbelt Trail that runs from Cold Spring Harbor in the north to Massapequa Preserve in the south.

The main entrance to Trail View State Park is located just east of a supermarket on Jericho Turnpike in Woodbury. Multiple trailheads provide access where Syosset-Woodbury Road meets Woodbury Road, and trails from Cold Spring Harbor State Park and Bethpage State Park lead to Trail View as well.

If you take the northward route from the main entrance, the hilly dirt trails pass through a small grove of Japanese maple trees that are the most astounding shade of red in the fall. The trail continues on the other side of Syosset-Woodbury Road, and if you cross the railroad tracks, you enter the 287-acre Stillwell Woods Park, where the trails continue. The larger western portion of Stillwell was once a functioning early twentieth-century sod farm operated by the Stilwell family, whose name was later misspelled. The Stilwell

The narrow Trail View State Park winds through Long Island's varying terrain from the North Shore to the south. *Author's collection.*

family lies buried in the small cemetery on Stillwell Lane. Nassau County acquired the farm for parkland in 1973. The Caumsett State Parkway right-of-way is on the eastern portion.

The trail crosses Stillwell Lane and continues north along Route 108, where it meets the Uplands Farm Sanctuary on its eastern side. The fifty-three-acre Uplands Farm, manor house, barn and outbuildings were gifted to the nature conservancy in the 1970s by Jane Nichols, the granddaughter of J. Pierpont Morgan, the infamous Gilded Age financier and investment banker. Caumsett State Parkway would have run right alongside Uplands Farm. Today, Cold Spring Harbor Laboratory operates a plant genetics lab on the site, and the property has been left in its pristine state, with many of the former farm buildings still standing.

North of Lawrence Hill Road, the Greenbelt Trails continue into Cold Spring Harbor State Park, which was also part of the Caumsett State Parkway right-of-way. That park is known for its dramatic hills and challenging hiking trails that overlook picturesque Cold Spring Harbor. The state dedicated it as a park in 2000.

These interconnected parks and trails run north to create an incredibly diverse and scenic byway.

From the main Trail View entrance on Jericho Turnpike in Woodbury, there are also trails that run south along the planned parkway to Bethpage State Park. The right-of-way continues, undeveloped, on the south side of Jericho Turnpike, in between housing developments and stores for a short distance to Woodbury Road. The undeveloped pathway continues to Northern State Parkway, where the state planned an exit and entrance ramp to Caumsett State Parkway. It continues south to the Long Island Expressway, where another ramp was planned, and alongside East Bethpage Road, where there is a parking lot near the intersection with Old Country Road so visitors can explore the trails there along the Bethpage Bikeway.

Crossing over Old Country Road, the right-of-way continues south, past Old Bethpage Road and Haypath Road, into Bethpage State Park. There, the planned parkway would have continued along the western side of what is now Bethpage Black Golf Course and connected to the existing portion of Bethpage State Parkway that leads to Southern State Parkway. On the south side of Southern State Parkway is the Peter J. Schmitt Massapequa Preserve, which runs down to Merrick Road. This preserve is part of the 2,200 acres the City of New York dedicated to the Long Island State Park Commission in 1925. The City of Brooklyn originally purchased the land, along with the land at Valley Steam State Park and Hempstead Lake State Park, to use for drinking water supply. Brooklyn later became part of New York City and no longer needed the water source. The transfer of these properties to the state preserved the ecologically important freshwater streams and lakes at Massapequa, Valley Stream and Hempstead. The Greenbelt Trails run through the 400 acres that surround Massapequa Lake and its creeks.

This sliver of land that runs from Cold Spring Harbor to Bethpage State Park and then connects with Massapequa Preserve completely bisects Long Island from north to south. Trail View offers a glimpse of the different types of terrain that existed in an earlier time on Long Island, from the low, flat plains and marshes of the South Shore up to the wooded and rocky highland terrain of the North Shore. The elevation ranges from sixty feet to three hundred feet above sea level.

The trails provide a peaceful, albeit narrow, course among the housing developments, busy roadways and strip malls that make up much of the Long Island landscape today. The park covers over four hundred acres and has fifteen miles of scenic wooded hiking and biking trails throughout. There are few facilities at Trail View State Park aside from small parking areas and map kiosks. Biking is permitted in paved areas, although many of the trails are suitable only for hiking. It is recommended to view trail maps on the state parks or Long Island Greenbelt Trail websites and to cross-reference regular roadmaps before navigating the trails, since some portions run along and cross over busy roadways.

VALLEY STREAM STATE PARK

Valley Stream State Park and the Village of Valley Stream are named after the freshwater streams that run through the small state park here. However, there are no nearby hills that would seem to situate it in a valley. The term *valley* is used on broader topographical maps to describe this part of Long Island as it contrasts with the rolling hills of the North Shore, located many miles away. The entire Valley Stream area is actually quite flat, and Natives who once lived here referred to it as "the plains country." The region was filled with vast meadows and grasslands, as well as bubbling streams of fresh water.

In the seventeenth century, Thomas and Christopher Foster settled a vast piece of land that encompassed the present-day Laurelton in Queens, Elmont, Valley Stream and Franklin Square for grazing purposes. The area was known as Foster's Meadow at that time. In the eighteenth and nineteenth centuries, a mix of Dutch and English farmers settled in the region, along with a significant German farming community in the mid-1800s.

Valley Stream was given its current name by Robert Pagan, who settled directly adjacent to the stream at the current state park. The Pagan family moved to the area in the 1840s and established a general store, post office and church. The fledgling town center was located in the area of Hendrickson Avenue, where that road bisects Valley Stream State Park and Arthur J. Hendrickson Park. Visitors to the area can get a feel for what life was like during that period by visiting the Pagan-Fletcher Restoration on the southwest corner of Valley Stream State Park. While Pagan is the

Valley Stream Water Works at Mill Road. *Valley Stream Historical Society.*

family's original name, Pagan's son John later changed the family name to Payan, because their neighbors thought their name sounded a bit impious, according to the Valley Stream Historical Society.

In the 1880s, the City of Brooklyn acquired the freshwater reservoirs at Valley Stream to serve as its drinking water supply. At that time, Brooklyn was an independent city separate from New York City, and it needed a clean water source to cater to its growing population. Brooklyn Water Works was responsible for damming the reservoirs and creating pumping stations to transport fresh water to the Ridgewood Reservoir, which was built in 1858.

Brooklyn also acquired what is now Hempstead Lake State Park for the same purposes. The city pumped the water through twelve miles of masonry conduits to the Ridgewood Reservoir, which still exists as a protected wetland in what is now part of a public park on the Brooklyn–Queens border called Highland Park. After Brooklyn merged with the City of New York in 1898, it gained access to the city's water supply from Upstate New York, which made it no longer reliant on the water supply from Nassau County.

In 1925, Robert Moses, the Long Island State Park commissioner, asked the City of New York to dedicate the unused streams, lakes and surrounding land in Nassau County to the park commission for public recreation. The City of New York agreed to turn over 2,200 acres of land in the form of a surface-use easement. The dedication included the present-day Hempstead Lake State Park, Valley Stream State Park, Massapequa Preserve and significant acreage for portions of the Southern State Parkway, Meadowbrook Parkway, Wantagh Parkway and Bethpage Parkway.

Valley Stream became a popular destination among visitors who came from New York City to enjoy the fields, streams, woodlands and lakes. There were various hotels and taverns situated around the area, primarily around Merrick Road. Later, the Park Commission promoted Valley Stream State Park as the nearest state park to New York City, located only two miles away from the Queens border. Crowds flocked there during the hot summer months.

A 1927 letter from State Park Chief Engineer A.E. Howland to the *Nassau Daily Review* reported on the progress of the new state park, which, at that time, also included the reservoir now located in the Village of Valley Stream's Arthur J. Hendrickson Park to the south of Hendrickson Avenue. He noted that a little over a month after receiving permission to use the reservoir for swimming, a bathhouse and concession stand were open. The bathhouse had 552 lockers, along with bathing suits, towels and rowboats available for visitors to rent. There were lifelines placed in the water for "inexpert swimmers," along with a "bathing crib" area for small children.

A postcard, circa 1930, showing the stream and pedestrian trail. Valley Stream State Park, Long Island, New York. *Boston Public Library, licensed under CC by 2.0.*

The state also erected a diving board at the site, and the park had picnic and camping areas.

Unfortunately, by 1948, the state was unable to maintain sufficiently clean water quality levels and discontinued swimming at the reservoir.

In 1958, the Village of Valley Stream purchased the twenty-five acres that contain the reservoir from the City of New York. While the reservoir was originally part of Valley Stream State Park, the state never fully owned it and used it only with permission from the City of New York. The Village of Valley Stream reportedly offered $50,000 to use the southern portion between Hendrickson Avenue and East Merrick Road as a village park. The pond is no longer used for swimming, but the village park has two swimming pools, tennis courts, basketball courts, a community center, a pavilion and hiking and biking paths around the reservoir. The park is named for a descendant of the early Hendrickson settler family, who negotiated with the state and the city to acquire the southern portion of the property for the village.

At Valley Stream State Park, the small freshwater "valley stream" trickles nearly forgotten through a wooded area on the east side of the park along Corona Avenue before it crosses under Hendrickson Avenue and feeds into the reservoir. Today, Valley Stream State Park is known for playing fields, a rentable pavilion, picnic areas with tables, fireplace grills, playgrounds and nature trails for hiking and biking. There is also a fitness loop with fifteen exercise stations. At ninety-seven acres, Valley Stream State Park is one of the smaller state parks in the system. However, the park offers residents and visitors the opportunity to get outside and enjoy the park amenities.

WILDWOOD STATE PARK

Wildwood State Park in Wading River used to be the site of two massive mansions that stood on the high bluffs overlooking the Long Island Sound during the early 1900s. One of the estates, built by famed New York City architect Stanford White, was located on the west side and was never lived in, since its owner died before it was completed. The other estate, built on the east side, was constructed by a character of such a dubious repute, his business dealings at the time were the subject of news headlines in the *New York Times*.

While the park is located in a scenic corner of Long Island's eastern half, it is far from the Gold Coast section of Long Island's North Shore, where Gilded Age millionaires typically built their grand summer mansions. However, a railroad line that was in operation between 1895 and 1938 brought more visitors to the once sparsely populated area. Many people from the elite class in New York City began to construct country houses to enjoy the peaceful surroundings and the beaches of the North Shore. For centuries before this period, the park was the site of the Hulse family homestead. The Hulses, along with other early settlers to the area, including the Lanes, Harods, Howells, Terrys, Youngs and Hortons, used the land for farming and grazing.

In 1905, Roland G. Mitchell came to the area and envisioned a secluded retreat far from the bustle of New York City. Roland Mitchell, also spelled Rowland, was a New York City financier and the principal of Roland G.

Roland G. Mitchell estate designed by famous New York Architect Stanford White. *Wading River Historical Society.*

Mitchell and Company, paraffin candle manufacturers. The candle company operated out of a building on First Avenue and Fourth Street in New York City amid the tenement buildings of that time. Mitchell used a proprietary method to create his candles, and the company did a brisk business, even as electricity was becoming more common. However, the factory had a string of bad luck over the years, with several fires reported at the building, including two that burned it down.

Mitchell must have been a fairly private person, as evidenced by the remote location where he chose to build his estate. There is very little mentioned of him in the society pages from the time, and he may not have come from money, since there's mention of him as an alumnus of the Free Academy, now called the City College of New York.

However, when Mitchell acquired the property, he had the money to employ one of the most prestigious Gilded Age architectural firms of the time: McKim, Mead and White. The firm's most recognized architect, Stanford White, worked on the design of the Mitchell mansion. McKim, Mead and White were the architects of such notable New York City structures as the Washington Square Arch, the original Penn Station, Bowery Savings Bank, the Prison Ship Martyrs' Monument and others. White was constantly in the society pages at the time and was known for his womanizing and fondness for showgirls. It was in the rooftop theater of the former Madison Square Garden, which he designed, where his lover's jealous husband shot and killed him during a performance in 1906, causing quite a scandal.

According to Stanford White's great-grandson Samuel White, the famous architect did some residential work on Long Island. Interestingly, he said, Stanford White was in the Wading River/Shoreham area a few years prior, designing the structure for famed inventor Nikola Tesla's laboratory and Wardenclyffe Tower, which was supposed to serve as a worldwide wireless communications apparatus a century ahead of its time. Samuel White noted:

> While the firm started out with a largely residential practice like any fledgling office, after the 1880s White was the only partner who regularly did both residential and commercial work, in addition to his institutional clients. He designed houses on Long Island for the very rich, for his family, and for his friends. He designed houses in Westbury for the Whitneys, and around Saint James for his sister-in-law and for himself. At the time of his death, he had just finished Harbor Hill, the house for Clarence and Katherine Duer Mackay in Roslyn, and a music room addition to the Orchard for James Breese, a friend in Southampton. Harbor Hill was torn down in the late 1940s, Breese's house has been transformed into the Whitefield Condominium.
>
> The Wading River house is close to the projects he was doing for Nikola Tesla, so there might be a connection between the two.

For Mitchell, White created a three-story mansion of brick and stone in what is described as the New England Federalist, or Neo-Georgian, style, according to *Long Island Country Houses and Their Architects, 1860–1940*. The brickwork on the exterior was more typical of New England homes and was complemented by the ornamental Corinthian columns and large portico that evoke Federalist styling. The house's main rooms were vast and open and contained high ceilings and fireplaces to warm against the chill of Long Island's North Shore winds.

The main floor contained only three oversized rooms that had mahogany and Flemish oak wood walls. The second floor contained five bedrooms with private baths for each. The third floor contained twelve bedrooms and had a large balcony overlooking the Long Island Sound. Mitchell also commissioned the famed New York City landscape architectural firm

Olmsted Brothers, known for designing grand outdoor spaces such as Fort Tryon Park in New York City, the National Cathedral in Washington, D.C., and other public and private works, to plan the grounds. The Olmsteds' father, Frederick Law Olmsted, was the creator of the American landscape design movement at the turn of the twentieth century, and he, along with Calvert Vaux, designed New York City's world-famous Central Park in 1857.

While the Mitchell estate must have been spectacular, it was never completed. Mitchell died in 1906, the same year his famous architect, Stanford White, was brutally murdered. Mitchell left the property to his brothers Arthur and Albert. However, it is not believed anyone ever occupied the unfinished mansion on a permanent basis. Although the estate was never officially called Wildwood while Mitchell was still alive, it is believed the family called it that for the beautiful stands of old-growth trees on its wild, windswept cliff.

The house remained closed for twenty years until Robert Moses, the head of the Long Island State Park Commission, spotted the estate on a trip to the North Fork. In 1925, Roland Mitchell's heirs agreed to sell 328 acres to the Long Island State Park Commission for $83,000, although Arthur Mitchell donated his share to the state. In 1926, the state acquired an additional 66 acres from Bernie Holdings Corporation, a New York City real estate developer who held hundreds of acres on the North Shore. The Mitchell mansion was demolished in the 1930s.

The state did not acquire the final acres on the eastern end of the park, which today contains hiking and biking trails, for another forty years. While the land there has been largely left in a natural, wooded state, that portion of the park was once the site of a mansion called Driftwood Manor. In 1906, New York City banking tycoon Joseph G. Robin commissioned the architectural firm Palmer and Hornbostel to design the estate in a grand Mediterranean villa style. The architects were well known for their design of the Queensborough Bridge and Williamsburg Bridge in New York City, along with many other public works.

The Robin mansion was two stories high and extended along the top of the bluff, with terraces and porches on either side overlooking the Long Island Sound. The main floor contained several large rooms, including an unusual elliptical, or oval-shaped, living room, according to *Long Island Country Houses and Their Architects, 1860–1940*. The estimated cost of the mansion was over $100,000 at the time.

However, it is questionable how Robin, if that was his real name, got the funding to build the house. Soon after the house was completed, Robin

Driftwood Manor, the Joseph G. Robin estate. *Long Island State Parks*.

became embroiled in a series of scandals involving the Hamilton, Northern, Washington and Riverside Savings Banks, which made for sensational news headlines at the time. In 1910, Robin was accused of manipulating the funds of the banks and falsifying records. Allegedly, under the guise of numerous shell corporations, he gained access to hundreds of thousands of dollars. His alleged misdeeds caused the collapse of Northern Savings Bank and the near ruin of Washington Savings Bank.

Interestingly, around the time the district attorney was preparing an indictment against him, the *New York Times* reported that Robin had himself committed to MacDonald's Sanitarium, a mental institution in Central Valley, New York, where he had himself declared "insane." Prosecutors were convinced the diagnosis was a ruse and prosecuted him anyway. He eventually pleaded guilty to grand larceny charges and served a little over a year on Blackwell Island before he received a pardon from the governor at the time.

Afterward, he spent several years fighting various legal battles to clear his name, which might not have even been Robin. His actual surname

The wall and gate that once surrounded Driftwood Manor still stands as a reminder of the past. *Author's collection.*

might have been Robinovitch. During the trial in 1911, prosecutors claimed that he was a Russian immigrant from Odessa, although when his alleged Robinovitch parents were brought in to testify that he was in fact their son, he denied knowing them. The newspapers went crazy over the story.

Driftwood Manor was placed up for auction in September 1912 to satisfy the liens of the mortgage holders, including Washington Savings Bank, one of the banks he allegedly swindled. The estate was sold to a representative of the banking department for the state, according to the *New York Times.* There were many owners over the years, including Thomas Clynes; the Morgenthay-Seixas Company for Edith Williams; Alfred Wagg, who used it as a summer home; and William Miller. It must be noted that the sale to Clynes might not have gone through, since that sale apparently occurred the year before the 1912 auction, according to news reports. Additionally, there is confusion about Miller's purchase of the property in 1929, since reports at the time said he purchased it from Joseph Robin. As with much of Robin's, or Robinovitch's, life, the real estate dealings concerning Driftwood Manor were complicated. The estate's final owner, before the state added the property to the park system, was Arthur G. Meyer. He made his millions dealing in textiles, according to reports at the time, and purchased Driftwood Manor around 1930. Meyer passed away in 1936; however, his wife, Margaret Meyer, continued to live in the house and entertain there well into the 1940s, according to local society pages. Mrs. Meyer agreed to sell the property and transfer it to the state upon her death in the 1960s. The mansion was knocked down in the 1980s.

There are few remnants of the mansions that once stood next to each other on the present state parkland. One of the former barns is now used for park maintenance. On the east side of the property, there is a concrete wall and gate, along with a small building believed to be the remains of a caretaker's cottage that were once a part of the former Driftwood estate.

Today, the public can enjoy the 767-acre park with 2 miles of shoreline and 11 miles of winding trails that weave through stands of native woodland atop the scenic cliffs. Visitors are permitted to fish and swim in the calm North Shore waters. The park also contains facilities for 240 tents and 80 trailers, as well as scenic picnic grounds, grills and playgrounds. Cabins are also available for rental during three seasons of the year.

NOTE FROM THE AUTHOR

This book covers the twenty-seven major Long Island State Parks. New York State operates numerous other preserves, golf courses and historic sites on Long Island, including Massapequa Preserve, Napeague State Park, Walt Whitman Birthplace State Historic Site, Robert Moses Pitch and Putt Course, Bethpage State Park Golf Courses, Sag Harbor State Golf Course, Sunken Meadow State Park Golf Course, Jones Beach Energy and Nature Center, Amsterdam Beach Preserve and Montauk Downs State Park Golf Course.

BIBLIOGRAPHY

Introduction

Bleyer, Bill. "Burden of Pataki's Expansion: State Parks Budget Flat Despite New Acquisitions." *Newsday*, February 28, 2010.

Campanella, Thomas. "Robert Moses and His Racist Parkway, Explained." Bloomberg. July 9, 2017. https://www.bloomberg.com/news/articles/2017-07-09/robert-moses-and-his-racist-parkway-explained.

Caro, Robert. *The Power Broker: Robert Moses and the Fall of New York*. New York: Vintage Books, 1975.

Ennis, Thomas. "Gilmore D. Clarke, 90, Is Dead; Designed Major Public Works." *New York Times*, August 10, 1982.

Goldberger, Paul. "Robert Moses, Master Builder, Is Dead at 92." *New York Times*, July 30, 1981.

Living New Deal. https://livingnewdeal.org/.

Martinl, Fay. "Long Island's Public Domain." *New York Times*, June 11, 1950.

National Park Service. "Robert Moses Biography." January 22, 2021. https://www.nps.gov/gate/learn/historyculture/robert-moses-biography.htm.

New York Preservation Archive Project. "Robert Moses." June 4, 2023. https://www.nypap.org/preservation-history/robert-moses/.

New York State Archives. "Robert Moses Collection." June 5, 2023. https://www.archives.nysed.gov/research/robert_moses_long_island_state_park_commissionhttps://www.archives.nysed.gov/research/robert_moses_long_island_state_park_commission.

New York Times. "Long Island Park Gets an Addition." March 27, 1958.

NYC Roads. "The Long Island Parkway System." December 12, 2023. http://www.nycroads.com/history/parkway/.

NYS Governor's Press Office. "Governor Hochul Announces NYS Parks Centennial Celebration." Press Release. August 7, 2023.

Schlichting, Kara Murphy. *New York Recentered: Building the Metropolis from the Shore*. Chicago, IL: University of Chicago Press, 2019.

World History.US. "Cars in the 1920s—The Early Automobile Industry." August 8, 2017. https://worldhistory.us/american-history/cars-in-the-1920s-the-early-automobile-industry.php.

Chapter 1

Bayard Cutting Arboretum. "The History of the Bayard Cutting Arboretum." May 6, 2023. https://bayardcuttingarboretum.com/about/.

Bayles, Thomas. "Old Estates at Oakdale." *Long Island Advance*, May 11, 1961.

Blakelock, Chester. *History of Long Island State Parks*. New York: Lewis Historical Publishing, 1959.

Friends of Connetquot. "The History of the South Side Sportsmen's Club." May 5, 2023. https://www.friendsofconnetquot.org/history.asp?name=William_Bayard_Cutting.

Islip Bulletin. "Cutting Arboretum was a Great Gift of Nature." March 21, 1974.

Long Island State Park Commission. *Long Island State Parks & Parkways Report*. 1972.

National Register of Historic Places. National Register Information System. June 6, 2018.

New York Times. "Family Here Since 1770." November 16, 1949.

———. "W.B. Cutting Dies on a Train." March 2, 1912.

Novick, Susan. "Baronial Charm and Treats." *New York Times*, November 8, 2009.

Wiecks, Kevin. "Bayard Cutting Arboretum." International Oak Society. August 15, 2020. https://www.internationaloaksociety.org/content/bayard-cutting-arboretum.

Chapter 2

Blakelock, Chester. *History of Long Island State Parks*. New York: Lewis Historical Publishing, 1959.

Bouvier, Jean. "Rothschild Family: European Family." *Encyclopaedia Britannica*. 2022. https://www.britannica.com/topic/Rothschild-family/additional-info#history.

Caro, Robert. *The Power Broker: Robert Moses and the Fall of New York*. New York: Vintage Books, 1975.

Drager, Marvin. "August Belmont: American Banker." *Encyclopaedia Britannica*. 2022. https://www.britannica.com/biography/August-Belmont.

Fleming, Rob. "Eating the Southern State." Small History Long Island. June 27, 2014. https://smallhistorylongisland.wordpress.com/2014/06/27/eating-the-southern-state/.

Gary, A. "August Belmont's Mansion." Home in Babylon. March 15, 2016. https://homeinbabylon.com/2016/03/15/august-belmonts-mansion/.

Miller, Tom. "The Lost August Belmont Mansion-No. 109 Fifth Ave." *Dayton in Manhattan*. April 18, 2016. https://daytoninmanhattan.blogspot.com/2016/04/the-lost-august-belmont-mansion-no-109.html.

Murphy Schlichting, Kara. *New York Recentered*. Chicago, IL: University of Chicago Press, 2019.

New York Times. "August Belmont Is Dead." November 25, 1890.

———. "State Gets Tract in Belmont Estate." April 2, 1926.

Chapter 3

Anderson, Dave. "U.S. Open at Bethpage, Five Courses and a Long History." *New York Times*, June 10, 2002.

Bailey, Paul. "Historic Long Island." *Suffolk County News*, February 12, 1954.

Blakelock, Chester. *History of Long Island State Parks*. New York: Lewis Historical Publishing, 1959.

Central Park Historical Society Encyclopedia. "Bethpage: From Settlement to the Early 20th Century." December 20, 2013. http://www.bethpagehistory.org/wiki/index.php?title=Bethpage.

———. "US Golf Open—2002." November 11, 2013. http://www.bethpagehistory.org/wiki/index.php?title=US_Golf_Open_-_2002.

Farmingdale Public Library. "Farmingdale Local History: Bethpage Purchase." May 16, 2023. https://farmingdalelibrary.libguides.com/c.php?g=743536&p=8092946.

Hammond, Gary. "Mystery Foto #27 Solved: Model of the Proposed Golf Courses of Bethpage State Park Golf (Circa 1933)." *Vanderbilt Cup Races*. July 6, 2015.

Historic American Buildings Survey. *Powell House, Farmingdale, Nassau County, NY.* Washington, D.C.: Library of Congress, 1933. https://www.loc.gov/item/ny0327/.

Orozco-Vallejo, Mary M. "Biography: Yoakum, Benjamin Franklin (1859–1929)." Texas State Historical Association. February 1, 1996. https://www.tshaonline.org/handbook/entries/yoakum-benjamin-franklin.

Starace, Carl A. "Historic Long Island." *Islip Bulletin*, May 8, 1975.

Wright, A.E. "Tywacana Farms Poultry Farmingdale." *Country Life in America*, 1913.

Chapter 4

Ferguson, Phillip M. "Thomas Story Kirkbride: American Psychiatrist." *Encyclopaedia Britannica*. December 12, 2022. https://www.britannica.com/biography/Thomas-Story-Kirkbride.

Murphy, Mardita. *The Kirkbride Plan: A Doctor's Interpretation of Moral Treatment and Its Reflection on Architecture.* UNCG Department of Interior Architecture: The Historic Dimension Series. Greensboro: University North Carolina, 2015.

New York Times. "Plan New State Hospital: Present Institutions Are Inadequate, Dr. Pilgrim Reports." December 20, 1916.

———. "State Is Building Biggest Hospital." May 26, 1929.

NYS Office of Mental Health. "Pilgrim Psychiatric Center-History." March 7, 2019. https://omh.ny.gov/omhweb/facilities/pgpc/.

Polaski, Leo. *The Farm Colonies: Caring for New York City's Mentally Ill in Long Island's State Hospitals.* Kings Park, NY: Kings Park Heritage Museum, 2003.

Stuhler, L.S. "Thomas Kirkbride Story 1809–1883—Physician, Psychiatrist and Developer of the Kirkbride Plan." Social Welfare History Project. May 25, 2023. https://socialwelfare.library.vcu.edu/organizations/state-institutions/kirkbride-thomas-story/.

Chapter 5

Argonne National Lab. "Reactors Designed by Argonne National Laboratory." October 7, 2019. https://www.ne.anl.gov/About/reactors/early-reactors.shtml.

Bayles, Donald, and Paul Infranco. *The History of Camp Upton: WWI Through WWII*. New York: Longwood Alliance Inc. and Longwood Society for Historic Preservation, 2017.

Bencivenga, Dominic. "Trailblazers Step by Step, Group Cutting a Shore-to-Shore Nature Path." *Newsday*, November 1, 1987.

Brookhaven National Lab. "BNL: Our History." www.bnl.gov.

Durr, Eric. "New York City Draftee Soldiers Made History as the Lost Battalion in October 1918." U.S. Army. October 2, 2018. https://www.army.mil/.

Longwood Society for Historic Preservation. "Historic Marker: Camp Upton." William G. Pomoroy Foundation. June 1, 2023. https://www.wgpfoundation.org/historic-markers/camp-upton/.

National Park Service. "Emory Upton: Antietam National Battlefield." May 30, 2023. https://www.nps.gov/people/emory-upton.htm.

New York Times. "Army School at Upton: Is First of Educational Centres to be Established in Camps." May 21, 1919.

———. "Camp Upton's 6,000 Acres to Be a State Park with Pine Forest and a Game Sanctuary." December 6, 1925.

Person, Gustav. "Union Blue and Militia Gray: The Role of the New York State Militia in the Civil War." Queens College of the City of New York. March 17, 2005. https://museum.dmna.ny.gov/unit-history/conflict/us-civil-war-1861-1865/union-blue-and-militia-gray-role-new-york-state-militia-civil-war.

University of Idaho. "Justice Department and U.S. Army Internment Camps and Detention Stations in the U.S. During World War II." College of Letters, Arts and Social Sciences, Alfred W. Bowers Laboratory of Anthropology. May 31, 2023. https://www.uidaho.edu/class/anthrolab/collections/aacc/research/kooskia/camps.

U.S. News & World Report. "In 1942, Business Owners Forced into Concentration Camps." December 5, 2016. https://www.usnews.com/news/national-news/articles/2016-12-05/from-the-archives-government-seizes-firms-puts-axis-citizens-in-concentration-camps.

Chapter 6

Friends of Caleb Smith Preserve. "History." February 27, 2019. www.friendsofcalebsmith.org/history.php.

Jacobson, Aileen. "From Elite Duck-Hunting Enclaves to Public Pastures." *New York Times*, July 29, 2011.

Newsday. "NY to Open Gates of Private Parks." April 28, 1973.

New York Times. "Preserves Build Up Supplies of Game." July 14, 1928.

————. "State Gets L.I. Club as Nature Preserve." June 7, 1963.

————. "Wyandanch Preserves: Plenty of Game for the Club Members at Smithtown." December 22, 1895.

Smithtown News. "State Airs New Master Plan." February 9. 1989.

Town of Smithtown. "History of Smithtown." June 5, 2023. www.smithtownny.gov/284/History-of-Smithtown.

Woodhaven Cultural and Historical Society. "The History Behind Woodhaven's Dexter Park: Our Neighborhood, the Way It Was." April 9, 2022. https://qns.com/2022/04/the-history-behind-woodhavens-dexter-park-our-neighborhood-the-way-it-was/.

Chapter 7

Draffen, Duayne. "Roosevelt's Rough Ride Led to Montauk." *New York Times*, May 17, 1998.

Fischer, William, Jr. "Antiaircraft Training at Camp Hero." Historical Marker Database. June 14, 2023. https://www.hmdb.org/m.asp?m=140310.

————. "Downtown Camp Hero." Historical Marker Database. June 14, 2023. https://www.hmdb.org/m.asp?m=140591.

Gilmartin, Richard T. "Montauk Naval Air Station, 1918." Montauk Library Archive. June 13, 2023. https://nyheritage.contentdm.oclc.org/digital/collection/p15281coll78/id/200/.

Kennedy, Dana. "Inside the Real-Life Lab and 'Secret Experiments' That Inspired *Stranger Things*." *NY Daily News*, October 24, 2020.

McSherry, Patrick. "Camp Wikoff (Montauk Point, Long Island, New York)." Spanish American War Centennial. June 13, 2023. https://www.spanamwar.com/campwikoff.html.

Navy Beach. "Montauk." June 12, 2023. https://www.navybeach.com/montauk/about-navy-beach/montauks-history/.

New York Times. "Plans Air Service for Coastal Defense: Admiral Usher Submits Outline of Navy's Needs in the Fifteen Districts." April 22, 1917.

Rosenbloom, Stephanie. "Under the Radar, a Montauk Park." *New York Times*, November 24, 2006.

Sheridan, Michael. "Secret Nazi Saboteurs Invaded Long Island during World War II, MI5 Documents Reveal." *NY Daily News*, April 4, 2011.

Speigel, Lee. "*Montauk Chronicles*' Claims Time Travel, Mind Control, Aliens at Camp Hero." *Huffington Post*, December 6, 2017.

Trowbridge, David J. "Camp Hero, Montauk Airforce Station." Clio: Your Guide to History. March 8, 2016. https://www.theclio.com/entry/21840.

Uboat.net. "Norness: Panamanian Motor Tanker." March 28, 2019. https://uboat.net/allies/merchants/1248.html.

United States Coast Guard Aviation History. "1917: Coast Guard Aviation in World War I." June 12, 2023. https://cgaviationhistory.org/1917-coast-guard-aviation-in-world-war-i/.

U.S. Army Corps of Engineers New England District. "Camp Hero FUDS, Montauk, New York." June 12, 2023. https://www.nae.usace.army.mil/Missions/Projects-Topics/Camp-Hero-FUDS-Montauk-New-York/.

Chapter 8

Blom, Arnold. "More Beach Space: Development of Captree Park to Improve Connections with Fire Island Shore." *New York Times*, October 18, 1953.

Caro, Robert. *The Power Broker: Robert Moses and the Fall of New York*. New York: Vintage Books, 1975.

Morris, Tom. *Island of Content: A History of Oak Island, Oak Island Beach and Captree Island, New York*. Babylon, NY: Town of Babylon, New York, Office of Historic Services, 2016.

———. "WaWa Yanda." *Newsday*, August 21, 1962.

Nack, William. "Island Retreat Is Pulled from the Past." *Newsday*, July 1, 1970.

New York Times. "Army Approves Project: Captree Causeway Across Great South Bay to Link Parkway." September 1, 1944. https://www.proquest.com/hnpnewyorktimes/docview/106825129/EEA2AEC3FD164472PQ/6?accountid=35174.

———. "Captree Span Complete." December 20, 1953.

———. "Coast Guard to Build on State Park Land." February 7, 1932.

———. "$5,609,540 Is Asked in Long Island Park Budget." November 22, 1930.

———. "Oak Beach Hotel Burns." October 6, 1924.

———. "Steel for Captree Span: Year's Work on Superstructure Will Begin in June." March 8, 1953.

O'Brian, Rachel. "Historic Coast Guard Station Damaged in Sandy Reopens in Oak Beach After $1.6M Renovation." *Newsday*, July 22, 2019.

Porterfield, Byron. "Fire Island Park a Ten-Minute Ride." *New York Times*, April 6, 1954.

South Side Signal. "Babylon Local Record: Death of Henry Havemeyer." June 5, 1886.

U.S. Coast Guard/Department of Homeland Security. "Station Oak Island Beach New York." U.S. Coast Guard History Program. June 21, 2023. https://media.defense.gov/2017/Jul/04/2001772873/-1/-1/0/OAKISLANDNY.PDF.

U.S. Lifesaving Service Heritage Association. "Oak Island Station House." June 21, 2023. https://uslife-savingservice.org/station-buildings/oak-island-station-house-2/.

Werden, Lincoln. "Wood, Field and Stream: A Report on State Fishing Large Sea Bass Taken to Revise Conservation Law." *New York Times*, June 19, 1937.

Chapter 9

Foreman, John. "A Damaged Beauty." *Big Old House*. March 21, 2013. https://bigoldhouses.blogspot.com/2013/03/a-damaged-beauty.html.

Hanc, John. "Historians Dig into Revolutionary War Past on Long Island." *Newsday*, July 23, 2019.

Handley, John. "Prairie Revival." *Chicago Tribune*, September 12, 2004. https://www.chicagotribune.com/news/ct-xpm-2004-09-12-0409110307-story.html.

Incorporated Village of Lloyd Neck. "Lloyd Neck: Early History." June 26, 2023. https://www.lloydharbor.org/about/.

Kolos, Walter D. "Caumsett, a Gold Coast Jewel." Caumsett Foundation. April 1, 2019. https://caumsett1.tripod.com/Caumsett_Site/GoldCoastJewel.html.

Lawrance, Gary. "Caumsett at Lloyd Neck, NY Estate of Marshall Field III." January 2, 2011. www.mansionsofthegildedage.com.

MacKay, Robert B., Anthony K. Baker and Carol A. Traynor. *Long Island Country Houses and Their Architects, 1860–1940*. New York: Society for the

Preservation of Long Island Antiquities in association with W.W. Norton & Co., 1997.

New York City Roads. "The Roads of Metro New York." March 11, 2019. http://www.nycroads.com/roads/.

New York Times. "Henry Field Dies in a Hospital Here." July 9, 1917.

———. "Marshall Field Dies at the Age of 63." November 6, 1956.

PBS: American Experience. "Chicago: City of the Century, Marshall Field (1834–1906)." June 27, 2023. https://www.pbs.org/wgbh/americanexperience/features/chicago-marshall-field-1834-1906/.

Wilson, Mark. "Field (Marshall) & Co." In *Encyclopedia of Chicago.* Chicago, IL: Chicago Historical Society, 2005.

Chapter 10

Atkinson, Baylee Browning. "A Woman Has an Awful Lot to Thank a Whale For." Cold Spring Harbor Whaling Museum. June 29, 2021. https://www.cshwhalingmuseum.org/blog/category/whaling-history.

Beers, F.W. *Long Island 1873: Cold Spring Town, Huntington Harbor Town, Deer Park Town, Northport Town.* Long Island, NY: Beers, Comstock and Cline, 1873. New York Public Library Digital Collections. https://digitalcollections.nypl.org/items/510d47e2-637a-a3d9-e040-e00a18064a99.

Charles Hosmer Morse Museum of American Art. "Laurelton Hall." July 5, 2023. https://www.morsemuseum.org/louis-comfort-tiffany/laurelton-hall.

Fresco, Robert. "The Past Is Present." *Newsday*, December 29, 1977.

Goldberger, Paul. "All World's Fair Houses Were But Fantasies of Everyday Life." *New York Times*, June 10, 1982.

Hughes, Robert. *Cold Spring Harbor.* Charleston, SC: Arcadia Publishing, 2014.

Kahn, Eve M. "Antiques: Resurrecting Laurelton Hall." *New York Times*, August 6, 2010.

Newsday, "Caumsett Plan Is Official." May 22, 1961.

New York City Roads. "Bethpage State Parkway." March 11, 2019. www.nycroads.com.

New York Times. "At Shady Glanada." July 29, 1894.

———. "A Summer Eden: Cold Spring Harbor Growing in Popularity as the Years Go By." May 27, 1893.

————. "Tiffany's Estate May Be a Rest Home." August 29, 1947.

Swift, Maurice. "Say New Road Planned to Link Caumsett Park." *Newsday*, May 5, 1961.

Chapter 11

Bergen, Gil. "The History of the South Side Sportsmen's Club." Friends of Connetquot River State Park Preserve. https://www.friendsofconnetquot.org/gil-bergen.

Eichel, Larry. "LI's Newest Parkland: The Aim Is to Preserve Their Tranquil Beauty." *Newsday*, May 15, 1973.

Friends of Connetquot. Tour of Connetquot State Park. On-site. July 12, 2023.

Historical Marker Database. "The Connetquot River." July 10, 2023. https://www.hmdb.org/m.asp?m=147232.

Karas, Nick. "The Old Mill Pond Is Still Great Fishing." *Newsday*, April 15, 1984.

MacKay, Robert B., Anthony K. Baker and Carol A. Traynor. *Long Island Country Houses and Their Architects, 1860–1940.* New York: Society for the Preservation of Long Island Antiquities in association with W.W. Norton & Co., 1997.

New York Times. "Cultivation of Trout: Fish Culture at the South Side Sportsmen's Club." March 15, 1885.

Raines, Jay D. "A Rich Legacy of the Great South Bay." *Fire Island Tide*, August 3, 2012.

Smith, Don. "Rich Men's Playground." *Newsday*, October 22, 1972.

Chapter 12

Binn, Sheldon. "Sea Gulls Patrol Long Island Shore at Abandoned Coast Guard Stations." *Newsday*, January 25, 1951.

Cavanaugh, Mike. "Gilgo Beach Timeline." Surfing Against Pollution. July 26, 2023. https://gilgo.com/mobile/gilgo-beach-history/.

Long Island State Park Commission. *Long Island State Parks & Parkways Report.* 1972.

New York Times. "Coast Guard to Build on State Park Land." February 7, 1932.

Noble, Dr. Dennis L. "A Legacy: The United States Life-Saving Service." U.S. Lifesaving Service Heritage Association. www.uslife-savingservice.org.

Southside Beacon. Advertisement. August 21, 1914, 8.

———. "Local News." August 10, 1901.

———. "Local News." August 31, 1901.

Suffolk County News. "Island News Notes." September 1, 1905.

———. "Island News Notes." May 24, 1918.

Tuomey, Douglas. "What's in a Name." 1960. http://longislandgenealogy.com/name/1.htm.

U.S. Coast Guard/Department of Homeland Security. "Station Gilgo." U.S. Coast Guard History Program. July 28, 2023. https://media.defense.gov/2017/Jul/03/2001772723/-1/-1/0/GILGO.PDF.

U.S. Lifesaving Service Heritage Association. "Gilgo Station House." July 28, 2023. https://uslife-savingservice.org/station-buildings/gilgo-station-house-2/.

Chapter 13

Friedman, Stefan C. "Boys' Club Still Stirs the Pot." *New York Post*, May 13, 2001.

Kennedy, Kathleen. "Exploring Hallock State Park and Preserve." Peconic Land Trust. August 5, 2021. https://peconiclandtrust.org/blog/exploring-hallock-state-park-and-preserve.

Kriete, Susan. "Guide to the Boys' Club of New York Records 1876–2002." New York Historical Society Museum and Library. July 24, 2023. https://dlib.nyu.edu/findingaids/html/nyhs/ms3000_boys_club_ny/bioghist.html.

New York State Office of Parks, Recreation and Historic Preservation. "Jamesport State Park Draft Master Plan." September 23, 2008.

Sabin, Charles. "Good Work by Boys' Club: Youngsters at William Carey Camp Show Effects of Training." *New York Times*, July 4, 1927.

Wines, Richard. *Defense of the Eagle*. Jamesport, NY: Hallockville Museum Farm, 2018.

———. "History of Hallockville: The Museum Farm, the Neighborhood, and the Surrounding Land Including Hallock State Park Preserve." Hallockville Museum Farm, July 23, 2008. https://hallockville.org/about/history/#tab-id-1.

Chapter 14

Blakelock, Chester. *History of Long Island State Parks*. New York: Lewis Historical Publishing, 1959.

Blidner, Rachelle. "Islip Has Come a Long Way in 335 Years." *Newsday*, November 21, 2018.

Encyclopaedia Britannica. "Richard Nicholl, English Governor." August 1, 2023. https://www.britannica.com/biography/Richard-Nicolls.

Find a Grave. "Edwin A. Johnson." August 1, 2023. https://www.findagrave.com/memorial/195798858/edwin-a-johnson.

Historical Society of the New York Courts. "William Nicholl 1657–1723." July 31, 2023. https://history.nycourts.gov/figure/william-nicoll/.

Morgan, Rosalinda. "East Islip, Suffolk County: Early History." Active Rain. September 19, 2018. https://activerain.com/blogsview/5276466/east-islip--suffolk-country---early-history.

Munkenbeck, George. "Islip History." Town of Islip. August 1, 2023. https://www.islipny.gov/community-and-services/explore-islip/islip-history.

New York State Office of Parks, Recreation and Historic Preservation. "Heckscher State Park." August 1, 2023. https://parks.ny.gov/parks/136/details.aspx.

New York Times. "August Heckscher Dies in Sleep at 92." April 27, 1941.

Norwood News. "Happenings." November 5, 1907.

South Side Signal. "Editorial: A Correction." June 10, 1899.

Spinzia, Ray. "Deer Range Farm." Old Long Island. April 25, 2011. http://www.oldlongisland.com/2011/04/deer-range-farm.html.

Suffolk County News. "A New York Murder: Aged Millionaire Shot by an Old Enemy." May 7, 1899.

———. "Romance at Great River: A Secret Marriage Causes Big Commotion." September 6, 1901.

Suffolk Weekly Times. "Gave Up Fortune for Lover." June 22, 1907.

Van Liew, B. "Historic and Natural Districts Inventory Form." NYS Division for Historic Preservation. 1979.

World Biographical Encyclopedia. "Richard Nicholl, Governor." August 1, 2023. https://prabook.com/web/richard.nicolls/3758530.

Chapter 15

Blakelock, Chester. *History of Long Island State Parks*. New York: Lewis Historical Publishing, 1959.

Caro, Robert. *The Power Broker: Robert Moses and the Fall of New York*. New York: Vintage Books, 1975.

Conry, Tara. "Ride the Carousel at Hempstead Lake." *Malverne-Lynbrook Patch*, June 30, 2011.

Governor's Office of Storm Recovery. "Focus Area: Hempstead Lake State Park." April 18, 2019. www.stormrecovery.ny.gov.

Helou, Paul. "Outlasting Influence: Long Island's Native American Culture." *New York Makers*, November 27, 2013.

Hewitt Woodmere Public Library. "Native Americans in the Rockaways." July 6, 2010. https://ftlh.blogspot.com/.

Kroessler, Jeffrey A. "Brooklyn's Thirst, Long Island's Water: Consolidation, Local Control, and the Aquifer." Lloyd Sealy Library, John Jay College of Criminal Justice. August 8, 2023. https://lihj.cc.stonybrook.edu/2011/articles/test-article/.

Long Island History Project. "The Last of the Hempstead Plains." September 26, 2022. https://www.longislandhistoryproject.org/the-last-of-the-hempstead-plains/.

New York Times. "Brooklyn Taxpayers Protesting." May 26, 1875.

———. "To Add 2,000 Acres to State's Parks." November 30, 1925.

NYS Governor's Press Office. "Governor Hochul Announces Completion of $47 Million Improvement Project at Hempstead State Park." June 27, 2023.

Roberts, Kenneth Lewis. *Oliver Wiswell*. Camden, ME: Down East Books, 1999.

Town of Hempstead. "History of the Town." April 16, 2019. www.hempsteadny.gov/.

Chapter 16

Backlund, Bruce. "Dismantling the Old Smith Meal Fish Factory in Promised Land, 1972." *East End Then*. January 14, 2017. https://eastendthen.wordpress.com/about/.

Blakelock, Chester. *History of Long Island State Parks*. New York: Lewis Historical Publishing, 1959.

Davis, Mark J. *American Experience: Mr. Miami Beach.* New York: Public Broadcasting Station, February 2, 1998. https://www.pbs.org/wgbh/americanexperience/features/miami-carl-and-jane-fisher/.

D'Mello, Judy. "How Entrepreneur Carl Fisher Reimagined Montauk." *Modern Luxury: Hamptons,* July 30, 2018.

First Super Speedway. "The Prest-O-Lite Story: Union Carbide Corporation, Linde Division." September 20, 2003. www.firstsuperspeedway.com.

Garrison, Virginia. "Remembering Who 'Discovered' Montauk." *East Hampton Patch*, December 1, 2010.

Maier, Thomas, and Steve Wick. "Lost Indian Lands/Shinnecocks and Montauketts Fight to Regain Areas Taken in Questionable Deals." *Newsday*, March 22, 1998.

New York Times. "Montauk Sold by Auction." October 23, 1879.

Penny, Larry. "The History of Hither Hills." *East Hampton Press*, May 10, 2022.

Piket, Casey. "Carl Fisher Discovers Miami Beach." *Magic City: Miami History*. December 20, 2015. https://miami-history.com/carl-fisher-discovers-miami-beach/.

Rattiner, Dan. "Our Amazing History: Austin Corbin." *Dan's Papers,* November 20, 2021. https://www.danspapers.com/2021/11/our-amazing-history-austin-corbin/.

Renner, Tom. "Montauk: A Land of One Man's Dream." *Newsday*, August 26, 1958.

Strong, John. "Editorial: More on How Montauk Was Parceled Out." *New York Times*, March 31, 1991.

Chapter 17

Blakelock, Chester. *History of Long Island State Parks.* New York: Lewis Historical Publishing, 1959.

Gorchov, Howard. "Photo Tour: John F. Kennedy Memorial Wildlife Sanctuary." *Massapequa Patch,* June 17, 2010.

Hanc, John. *Jones Beach: An Illustrated History.* Essex, CT: Globe Pequot Press, 2007.

James, Birdsall. "A Personal Account." *Wantagh Preservation Society Newsletter* (May 2004). https://www.wantagh.li/museum/information_window/2004_05_information_window.pdf.

Lewin, Jonathan. "How 'Master Builder' Robert Moses Spearheaded the Creation of Long Island's Jones Beach." *Daily News*, August 14, 2017.

Long Island Newsday Time Machine. "Living the Good Life by the Sea." October 31, 1999.

Long Island Traditions. "Eastern Nassau." August 22, 2023. https://longislandtraditions.org/eastern-nassau/.

Rabon, John. "A Brief History of British Privateers and Pirates." Anglotopia for Anglophiles, May 5, 2021. https://anglotopia.net.

Rather, John. "King of Beaches Gets a New Title." *New York Times*, February 6, 2005.

Saslowmay, Linda. "Jones Beach Tower Set for Restoration." *New York Times*, May 18, 2008.

Schwab, Fred. "What About That Name…High Hill." High Hill Striper Club. January 19, 2024. www.highhillstriperclub.com.

South Side Messenger. "Bellmore: Robert T. Willmarth." September 1, 1911.

United States Coast Guard. "Station Short Beach: New York." August 12, 2023. https://www.history.uscg.mil/.

Wilkinson, Annie. "The Haunted History of Maj. Thomas Jones' 'Old Brick House' at Massapequa Manor." *Long Island Press*, October 17, 2022.

Chapter 18

Blakelock, Chester. *History of Long Island State Parks.* New York: Lewis Historical Publishing, 1959.

Caro, Robert A., and Wiemer, Robert. "State Plans 2 Big Parks in Suffolk." *Newsday*, July 30, 1962.

Chinese, Vera. "Montauk's Historic Third House, Where Teddy Roosevelt Slept." *Newsday*, August 10, 2020.

Conry, Tara. "Secrets of the Montauk Lighthouse." *Newsday*, March 31, 2019.

Friends of Montauk Downs. "Montauk Downs: The Course: History." May 14, 2019. www.montaukdowns.org.

Matouwac Research Center. "Early European Contact Years." December 12, 2023. http://www.montaukwarrior.info/?page_id=11.

Montauk Historical Society. "Montauk History Overview." August 30, 2023.

Montauk Library. "Montauk Historic Facts." August 30, 2023.

Penberthy, Bryan. "History of the Montauk Point Lighthouse." U.S. Lighthouses. January 1, 2016. https://www.us-lighthouses.com/montauk-point-lighthouse#google_vignette.

Rattiner, Dan. "Our Amazing History: Austin Corbin." *Dan's Papers*. November 20, 2021. https://www.danspapers.com/2021/11/our-amazing-history-austin-corbin/.

Rees Jones Inc. "Montauk Downs." May 14, 2019. https://reesjonesinc.com/courses/montauk-downs/.

Tikkanen, Amy. "Montauk People." *Encyclopaedia Britannica*, August 20, 2023. https://www.britannica.com/topic/Montauk-people.

Town of East Hampton LWRP. "Section VIII. Historic Resources Policy #23." May 14, 2019.

Werkmeister, Joe. "Where Native Tribes Once Warred, Expansion of Historic Montauk Cemetery Could be in the Works." *Newsday*, June 13, 2023.

Chapter 19

Allen, Lewis Falley. *The American Shorthorn Herd Book*. Kansas City, MO: American Shorthorn Breeders Association, 1872.

Flynn, David. *Early Houses of Kings Park*. New York: Kings Park Heritage Museum, 2004.

Harris, Brad. "News of Long Ago." *Smithtown News*, May 14, 1981.

Kings Park Heritage Museum Archives.

Long Islander. Legal notice. March 26, 1880.

Medina, Jason. *Kings Park Psychiatric Center: A Journey Through History*. Vol. 1. New York: Xlibris, 2018.

Native Long Island. "Nissequogue." August 14, 2020. https://nativelongisland.com/listing/nissaquogue/.

Polaski, Leo. *The Farm Colonies*. New York: Kings Park Heritage Museum, 2003.

The Wealth and Biography of the Wealthy Citizens of the City of New York. 10th ed. New York: New York Sun, 1846.

Chapter 20

Blakelock, Chester. *History of Long Island State Parks*. New York: Lewis Historical Publishing, 1959.

Clemente, T.J. "Orient Point: George Washington, Benedict Arnold, and the British Fleet in the Hamptons." Hamptons.com, May 29, 2013. https://hamptons.com/lifestyle-on-the-water-18250-orient-point-george-washington-benedict-html/.

Corrector. "Hon. Lewis A. Edwards." June 14, 1869.

Cottral, Dr. George E. "History of Orient." *Historical Review* (July 1959): 5–19.

Malone, Mary. "Get Oriented with Orient Point." *Dan's Papers.* February 27, 2022. https://www.danspapers.com/2022/02/get-oriented-with-orient-point/.

Merriam-Webster. "Orient." December 19, 2022. https://www.merriam-webster.com/dictionary/orient.

Taylor, David. "East End Seaport Museum: Keepers of Bug Light & Greenport History." *Dan's Papers.* August 30, 2022. https://www.danspapers.com/2022/08/east-end-seaport-museum-bug-light-visit/.

Wacker, Ronnie. "Fate of Historic Orient Inn Hangs in Balance" *New York Times,* July 1, 1984.

Young, Rachel. "A Once-Grand Hotel in Orient." *Northforker,* October 21, 2015.

Chapter 21

Andriotis, Elizabeth. "The Fascinating History of an Elsie de Wolfe–Designed Tea House." *House Beautiful*, April 16, 2021.

Cultural Landscape Foundation. "Olmsted: Planting Fields Arboretum State Historic Park." September 12, 2023. https://www.tclf.org/sites/default/files/microsites/landslide2022/locations/plantingfields.html.

Farmingdale Post. "W.R. Coe, Planting Fields Owner, Succumbs in the South." March 24, 1955.

Foreman, John. "Long Island People." *Big Old Houses*, October 8, 2014. http://bigoldhouses.blogspot.com/.

Joyce, Henry. *Guidebook to Coe Hall.* Oyster Bay, NY: Planting Fields Arboretum State Historic Park, 2023. http://www.macomeadesign.com/.

MacKay, Robert B., Anthony K. Baker and Carol A. Traynor. *Long Island Country Houses and Their Architects, 1860–1940.* New York: Society for the Preservation of Long Island Antiquities in association with W.W. Norton & Co., 1997.

New York Times. "Coe Home Burned with $700,000 Loss." March 28, 1918.
———. "Obituary: Mrs. William R. Coe." October 13, 1960.
———. "W.R. Coe, Sportsman, Weds Suddenly at 57." December 5, 1926.
O'Donnell, Patricia M. "Bringing the Planting Fields Cultural Landscape Report to the Ground: Main Drive Renewal Project." Heritage Landscape LLC, September 20, 2023. https://home.nps.gov/subjects/ncptt/upload/FG_Wouters-ODonnell.pdf.
Stern, Michael. "Oyster Bay Arboretum to Close." *New York Times*, July 4, 1971.
Stony Brook University Libraries. "S.B.U. History and Timeline." Special Collection and University Archives. September 10, 2023. https://guides.library.stonybrook.edu/sbuhistory.

Chapter 22

Blakelock, Chester. *History of Long Island State Parks*. New York: Lewis Historical Publishing, 1959.
Blom, Arnold, "More Beach Space: Development of Captree Park to Improve Connections with Fire Island Shore." *New York Times*, October 18, 1953.
Brevig, Walt. "Moses Bridge, Park Open; 35,000 Visit Over Weekend: So They Drove to Fire Island." *Newsday*, June 15, 1964.
Caro, Robert. *The Power Broker: Robert Moses and the Fall of New York*. New York: Vintage Books, 1975.
Columbia University Archives. "Robert Moses: If the Ends Don't Justify the Means, What Does?" September 30, 2023. https://c250.columbia.edu/c250_celebrates/remarkable_columbians/robert_moses.html.
Fire Island Travel Guide. "Fire Island History." September 30, 2023. https://fireisland.com/fire-island-history/.
Forman, Seth, and Lee Koppelman. *The Fire Island National Seashore: A History*. New York: SUNY Press, 2008.
Goldberger, Paul. "Robert Moses, Master Builder, Is Dead at 92." *New York Times*, July 30, 1981.
National Parks Service. "Origin of Fire Island." September 30, 2023. https://www.nps.gov/fiis/learn/historyculture/fireislandorigin.htm.
Porterfield, Byron. "Fire Island Park a 10-Minute Ride: Ferries from Captree, at East End of Jones Beach, Will Start Running by July 1." *New York Times*, April 6, 1954.

———. "$10 Million Fire Island Bridge to Be Opened to Public June 13: Span and State Park Named for Robert Moses." *New York Times*, June 4, 1964.

Silver, Roy. "About Long Island's Sea Gulls and State Parks." *New York Times*, May 26, 1968.

United States Coast Guard. "Fire Island Lighthouse." September 29, 2023. https://www.history.uscg.mil/Browse-by-Topic/Assets/Land/All/Article/1916122/fire-isla nd-lighthouse/.

———. "Life-Saving Service & Coast Guard Stations." September 29, 2023. https://www.history.uscg.mil/Browse-by-Topic/Assets/Land/Stations-Units/Article/2648559/station-fire-island-new-york/.

Wilkinson, Annie. "How the 1892 Cholera Pandemic Led to a Showdown at the Surf Hotel on Fire Island." *Long Island Press*, April 21, 2020.

Chapter 23

Bleyer, Bill. "Pataki's State Parks Expansion Added Budget Burden." *Newsday*, February 27, 2010.

Hanc, John. "Confronting a Crisis: How LI United to Fight an Epidemic in 1898." *Newsday*, April 3, 2020.

Heatley, Jeff. "Camp Wikoff, Quarantine Camp, Montauk." *Art & Architecture Quarterly, East End* (September 1898). https://aaqeastend.com/contents/camp-wikoff-col-theodore-roosevelt-the-rough-riders-sept-1-nov-9-1898/.

Nature Conservancy. "Discover Striking, Panoramic Views at this Beachfront Property." October 4, 2023. https://www.nature.org/en-us/get-involved/how-to-help/places-we-protect/long-island-shadmoor-preserve/.

Osmers, Henry. "Lost Montauk." *Montauk Historical Society Beacon*, May 23, 2023.

Powers Porco, Joan. *Holding Back the Tide: The Thirty-Five-Year Struggle to Save Montauk*. New York: Harbor Electronic Publishing, 2005.

Rather, John. "After 20 Year Effort Shadmoor's a State Park." *New York Times*, October 22, 2000.

Rattiner, Dan. "Our Amazing History: Teddy Roosevelt at Montauk." *Dan's Papers*, July 15, 2021. https://www.danspapers.com/2021/07/our-amazing-history-teddy-roosevelt-at-montauk/.

Strong, John. *The Montaukett Indians of Eastern Long Island*. New York: Syracuse University Press, 2001.

United States Coast Guard/U.S. Department of Homeland Security. "Station Ditch Plain, New York." October 5, 2023. https://www.history. uscg.mil/.

Chapter 24

Bleyer, Bill. "Sunken Meadow Park Renamed." *Newsday*, September 2, 1992.

———. "Sunken Meadow's 22-Acre Addition." *Newsday*, February 15, 2002.

Flanagan, Kerriann. "The Stately Platt Homestead in Fort Salonga." *Long Islander*, June 20, 1991.

Flynn, David. *Early Houses of Kings Park*. New York: Official Offset Printing, 2004.

Harris, Bradley, and King Pedlar. *St. Johnland, A Forgotten Utopia*. New York: Official Offset Printing, 2008.

Kings Park Heritage Museum Archives. Accessed 2020–2024.

Lader, Phyllis. "Show House Setting Rich in History." *Observer*, September 5, 1991.

Long Island State Park Commission. *Report of the Long Island State Park Commission*. 1930.

Long Island State Parks and Parkways. North Babylon, NY: Long Island State Parks Commission, April 1972.

New York Times. "Long Island Park Gets an Addition." March 27, 1958.

Patchogue Advance. "Homestead Added to Sunken Meadow Park." May 16, 1930.

Slayton, Robert A. *Empire Statesman: The Rise and Redemption of Al Smith*. New York: Free Press, 2001.

Tiernan, George Washington. *Kings Park: A Pictoral History*. New York: Official Offset Printing, 2007.

Chapter 25

Barone, John. "Editorial: The Tale of Caumsett. This Time, NIMBYers Knew More." *Newsday*, August 19, 2023.

Blakelock, Chester. *History of Long Island State Parks*. New York: Lewis Historical Publishing, 1959.

Bleyer, Bill. "Making Tracks/Land Once Slated for Parkway Becomes State Park for Hikers." *Newsday*, September 21, 2002.

Bunch, William. "Albany Awash in Red Ink: A Red Sea in New York State Fiscal Crisis Brings Back Memories of Mid-'70s." *Newsday*, March 12, 1990.

Freeman, Ira Henry. "Bikes Now-Polo Ponies Then: Once Upon a Time at Caumsett." *New York Times*, March 27, 1977.

Hanc, John. "Hidden Paradise: The New Trail View State Park Captures the Sights and Sounds of Pre-Suburban Long Island." *Newsday*, October 23, 2002.

Lambert, Bruce. "Nassau Considers Selling Park Land." *Newsday*, August 22, 1978.

Levine, Ned. "Why Isn't There an Exit 47 on the Long Island Expressway?" *Newsday*, September 13, 2023.

Newsday. "A View from Stillwell Woods." October 18, 1978.

Swift, Maurice. "Say New Road Planned to Link Caumsett Park." *Newsday*, May 5, 1961.

Chapter 26

Blakelock, Chester. *History of Long Island State Parks*. New York: Lewis Historical Publishing, 1959.

Caro, Robert. *The Power Broker: Robert Moses and the Fall of New York*. New York: Vintage Books, 1975.

Cerra, Francis. "Blazing a New Trail in the Wilds of Nassau." *New York Times: Long Island Weekly*, January 18, 1987.

Foster Meadow Heritage Center. "A German Farming Community, Est. 1850s on Long Island." January 8, 2024. https://www.fostersmeadow.com/.

Howland, A.E. "Letter to the Editor." *Nassau Daily Review*, August 25, 1927.

Martinl, Fay. "Long Island's Public Domain." *New York Times*, June 11, 1950.

Newsday. "NY Officials Expect Agreement on Park Land Sale to V. Stream." January 7, 1957.

New York City Department of Parks and Recreation. "Ridgewood Reservoir: A Brief History." October 26, 2023. https://www.nycgovparks.org/parks/highland-park/highlights/19651.

Ruehl, Howard. *50ᵗʰ Anniversary History of Valley Stream, 1840–1975*. New York: Incorporated Village of Valley Stream, 1975.

Schneps Media. "Queens' Early History Shaped by Indians." July 9, 2003. https://qns.com/2003/07/queens-early-history-shaped-by-indians/.

Village of Valley Stream. "History of Valley Stream." October 25, 2023. https://www.vsvny.org/.

Young, Michelle. "Ridgewood Reservoir Named on the National Register of Historic Places." Untapped New York. October 26, 2023. https://untappedcities.com/2018/02/09/ridgewood-reservoir-named-on-the-national-register-of-historic-places/.

Chapter 27

Alcorn, Jane, and Mary Ann Oberdorf. *Shoreham and Wading River.* Charleston, SC: Arcadia Publishing, 2012.

Blakelock, Chester. *History of Long Island State Parks.* New York: Lewis Historical Publishing, 1959.

Caro, Robert A. *The Power Broker: Robert Moses and the Fall of New York.* New York: Knopf, 1974.

County Review. "Roland G. Mitchell." June 22, 1906.

———. "Wading River Estate Sold to A.H. Wagg of Florida." July 23, 1925.

East Hampton Star. "State Park for Riverhead." January 23, 1925.

Larsen, Ray. "The Hidden Past of Wildwood State Park." Long Island Parks. November 5, 2023. https://liparks.com/.

Long Islander. "Suffolk Association Approves Two Parks." April 23, 1926.

MacKay, Robert B., Anthony K. Baker and Carol A. Traynor. *Long Island Country Houses and Their Architects, 1860–1940.* New York: Society for the Preservation of Long Island Antiquities in association with W.W. Norton & Co., 1997.

Nellen, A. Robert. *History of Wildwood State Park.* Wildwood State Park. N.d.

New York Times. "Cheney Shuts Northern Bank." December 28, 1910.

———. "Demands $2,750,000 of Bank's Lawyers." January 16, 1914.

———. "Flames in a Candle Factory." July 5, 1882.

———. "Guilt and Sane Is Robin's Plea." March 2, 1911.

———. "Pardon to Robin Issued by Sulzer." September 3, 1913.

———. "The Real Estate Field: Driftwood Manor, Country Estate of Joseph G. Robin, Bought in by State for $42,000." September 4, 1912.

Port Jefferson Echo. "Items of Interest in the Building Trade." May 9, 1929.

Roth, Leland M. *The Architecture of McKim, Mead & White: 1870–1920. A Building List.* New York: Garland Publishing, 1978.

Suffolk County News. "Island News Notes." November 3, 1911.

Suffolk Times. "State Park for Riverhead Town." January 16, 1925.

SPECIAL THANKS

Aimee Lusty, archivist, Montauk Public Library

Caren Zatyk, librarian, Richard H. Handley Collection of Long Island Americana (Long Island Room), Smithtown Library

Carol McKenna, Valley Stream Historical Society

Carol Poulos, curator, Wantagh Preservation Society

Chryssa Golding, reference and local history librarian, Bellmore Memorial Library

Cinda Lawrence, assistant to the president, Boys' Club of New York

Erik Huber, archivist, Queens Public Library

George Munkenbeck, town historian, Islip

Jane Alcorn, Wading River Historical Society

John Williams, Long Island State Parks

Karen Martin, Huntington Historical Society

Kings Park Heritage Museum

Mary Cascone, town historian, Town of Babylon Office of Historic Services

Melanie Cardone-Leathers, Longwood Public Library/Thomas R. Bayles Local History Room

Melissa Andruski, Southold Free Library

Peter Ward, local history librarian, Brentwood Public Library

Public Libraries of Suffolk County, Live-brary.com

Ray Lembo, chairman, East Islip Historical Society

Robert Hughes, town historian, Huntington

Samuel White, architect, PBDW Architects

ABOUT THE AUTHOR

Kristen Matejka is a longtime writer who has held various editorial positions at both local and national newspapers and magazines. She served as a marketing director at Discover Long Island, the Long Island Visitor's Bureau, for twelve years. During her tenure, she wrote countless narratives about places to visit and things to see on Long Island, covering a vast array of topics, including Long Island's Gold Coast, historic lighthouses and windmills, wineries, science and aviation history, the Revolutionary War, famous film sites and more. She also worked on a special Path Through History project for Long Island as part of New York State's Path Through History program.

She has a wealth of experience working for historical organizations on Long Island, where she shares fascinating stories about some of the most interesting places to visit. These include Southold Historical Museum on the North Fork and the North Shore Promotion Alliance's historic Washington Spy Trail initiative. She also volunteered as an archivist at the Bohemian Citizens Benevolent Society of Astoria, a Czech cultural organization, and serves as a trustee at Kings Park Heritage Museum, as well as on the Town of Smithtown Historic Advisory Board.

Visit us at
www.historypress.com
....